From Pioneering to Persevering

Indiana's Counties as of 1850

From Pioneering to Persevering

FAMILY FARMING IN INDIANA TO 1880

Paul Salstrom

Purdue University Press
West Lafayette, Indiana

Printed in the United States of America.

Grateful acknowledgment is made to use the following images:

Cover illustration: Watercolor of Vigo County scene in late 1840s painted by Sister Mary Magdelene, SP. Courtesy of the Sisters of Providence Archives, St. Mary-of-the-Woods, Indiana.

Emigrant Boat in which the Pioneers Went from Pittsburg to Kentucky, p. 40, from Bass Photo Company Collection, courtesy of Indiana Historical Society.

Dates of Indian Land Cessions in Indiana, p. 61, from Gregory S. Rose, "Hoosier Origins," *Indiana Magazine of History*, vol. 81 (Sept. 1985), p. 221. Used with permission.

Threshing machine, Wagon, and Farmers in Indiana in 1912, p. 84, from Bass Photo Company Collection, courtesy of Indiana Historical Society.

Solon Robinson, 1841, p. 94, courtesy of Indiana Historical Bureau as reproduced from *Solon Robinson: Pioneer Agriculturalist* (Indianapolis: Indiana Historical Bureau, 1936), vol. 1, frontispiece.

ISBN 978-1-55753-453-8

Library of Congress Cataloging-in-Publication Data

Salstrom, Paul, 1940-

From pioneering to persevering : family farming in Indiana to 1880 /
Paul Salstrom.

p. cm.

Includes bibliographical references and index.

ISBN 978-1-55753-453-8 (alk. paper)

1. Farm life--Indiana--History. 2. Family farms--Indiana--History.
3. Agriculture--Indiana--History. I. Title.

S521.5.I6S35 2007

630.9772--dc22

2007001056

Dedicated to the Memory of
Robert M. Taylor, Jr.

Contents

Illustrations

Preface

A geyser of emotional affection for Indiana family farms gushed from Hoosier ground sixty years ago when the Indiana Historical Society announced a "Centennial Farms" project. From 1947 to 1951, over 1,650 farm owners carefully (and successfully) documented that the farm they owned had been continuously owned by members of their family for over 100 years. So much documentation and explication went into those farm owners' reports that they are now a treasure trove for genealogists and local historians. (*Centennial Farms of Indiana* by Teresa Baer, et al. tells the story with the help of numerous photos.)

Then thirty years later, in 1976, the State of Indiana launched the Hoosier Homestead Project, which has led to another treasure trove of farm heritage.

So Hoosiers still care about family farming. The genesis of the present book was a letter that I wrote twelve years ago to a man much beloved by his friends, the late Bob Taylor, Jr. of the Indiana Historical Society. I asked him if the IHS might want to authorize the writing of a "new rural history" about rural Indiana. Bob Taylor, as it turned out, not only knew all about the new rural history but was a friend of its first proponent, Robert Swierenga. The plan became that I would write a history of Indiana agriculture (which is something that the IHS has long wished to make available) and in the process I would interject the kinds of grassroots social and economic questions that the new rural history tries to answer. That was the plan—but into it steadily crept the fate of easy-entry family farming until the question "why?" now dominates the latter part of the book.

Eventually I came to see that what really sealed the fate of most Indiana family farms was the introduction of mechanical corn harvesting. And when I saw that, I decided to wrap up the book rather than try to trace all the repercussions that family farming has consequently faced in the 1900s, or to trace all the splinters of family farming that still survive today.

But as I wrap up the project, I also find myself wondering why my version of Indiana's early agricultural history does not simply highlight the family-farming story—why it instead keeps asking *why.* Specifically, why family farming was so *economically viable* in early Indiana. Now, after tossing retrospective questions at family farming all through this book, I guess it is only fitting that I start wondering something about myself: why economic viability so dominates my questions.

Perhaps this is because my mentor in life was a radical carpenter named Bob Swann, who pointed me toward grassroots economic questions. He founded the community land trust movement in the 1960s and then later the E. F. Schumacher Society in Massachusetts to try to apply the ideas in Schumacher's book *Small Is Beautiful.* (Stephanie Mills is preparing a biography of Swann.)

Later, when I began to study history, I found myself putting grassroots economic questions to the history of Appalachia—a part of the U.S. where I had meanwhile been trying my hand at small-scale farming. (Incidentally, if Appalachia is broadly defined as the Upland South, that is where most of Indiana's early settlers arrived from.)

But so much for where this book came from. In fact, scholarship is a community enterprise, even if only one person is labeled the author. My thanks go not only to the Indiana Historical Society for supporting the writing of this book with a Special Project Grant but also to the Society's librarians, especially Wilma L. Gibbs. Thanks also to the staff of the Indiana Division of the Indiana State Library; and the staff of Indiana State University Library, especially to David Vancil and Dennis Vetrovec of Special Collections there. And the unflagging help given me by the staff of Rooney Library at my own college, Saint Mary-of-the-Woods College near Terre Haute, has made this book measurably better.

Others at my college have helped as well. The naturalist Marion Jackson carefully read the book's first half. Kim DisPennett, Lynn Hughes, and Sue Weatherwax helped me with the maps.

Finally, the expertise of Margaret Hunt, Rebecca Corbin, and Bryan Shaffer at Purdue University Press has also made the book better, and has made it look inviting.

From Pioneering to Persevering

Introduction

THE WOMEN HAD formidable skills of housekeeping and family management, and the men generally knew something of carpentry, construction, animal care, and simple machinery. . . . Never perhaps in the history of the world have nominally free and prosperous farmers, businessmen, and laborers worked so hard, such long hours, so energetically and with such visible reward.

—William N. Parker, "Native Origins of Modern Industry," pp. 246, 253

When Thomas Jefferson opened the Declaration of Independence with a nod toward world public opinion, voicing "a decent respect to the opinions of mankind," his concern did not likely extend beyond the opinions of Europeans.

But the years saw many changes, and when the Midwestern-born historian William N. Parker (quoted above) wrote about the mid-1800s in the Midwest that "never perhaps in the history of the world have nominally free and prosperous farmers, businessmen, and laborers worked so hard, such long hours, so energetically and with such visible reward," he really was comparing the Midwest with the entire rest of the world.

Unlike Thomas Jefferson, William Parker was not writing a Declaration. He simply hoped to explain how the Midwest had grown so successful. He did not feel compelled to justify it.[1] Parker had been born in 1919, the year when

Woodrow Wilson went to the Paris Peace Conference and tried to save the world from itself. Parker and his generation of young Americans then came of age in World War II and rode the wave of U.S. victory.[2] By that time the Midwest's agricultural output was soaring to levels able to feed a sizable portion of the world's people. In 1950, with 27 percent of its cropland devoted to corn, the U.S. was producing almost half of the corn grown worldwide.[3]

Massive food exports from the Midwest had already started by the time of the Civil War, and they depended on the money, shipping, and mass markets of the northeastern states.[4] By 1860, the Midwest no longer depended on shipping its products down the Mississippi River to New Orleans. By then its ties to the markets and money of the northeastern United States had become quite cozy.[5] Midwestern farms and factories helped the Union defeat the Confederacy, and then, unlike Southerners, Midwesterners didn't look backward and brood but just continued churning out more. "Why?" was not a discomfiting question in those days. Midwesterners were working like maniacs for reasons that to them seemed self-evident: for their livelihoods, their self-respect, and their self-betterment. (This latter was also lauded as "improvement" or progress, and it charmed more northerners than southerners.) Overall, most Midwesterners were land-hungry Euro-Americans, some were land-hungry African-Americans, and a few were persevering Native Americans.

This book will start in Chapters One and Two with some information about the agriculture of the indigenous people who found themselves displaced by all this hyperactivity, Indiana's Native Americans. Their agriculture was managed by women and they used no draft animals. Then Chapter Three will glance at the 1700s French settlers and Chapter Four at the early 1800s influx of settlers from Kentucky and the rest of the Upper South. After glancing in Chapter Five at the southern origins of the corn belt system, and in Chapter Six at the settling of central Indiana, we'll begin encountering Indiana's *other* main flow of early settlers—those from the northeastern states of Pennsylvania, New York, New Jersey, and the New England states—who started arriving in large numbers in the 1830s. Those settlers often reached Indiana after tarrying a few years or decades in Ohio (as also did many of Indiana's southern in-migrants). And we'll encounter still more incoming southerners and northerners in Chapters Seven and Eight, which focus on the settlement of western and northern Indiana respectively.

A pair of hard-working historians have determined that migrants from the Northeast often "were attracted to the new region by prospects of becoming more profitable farmers, bigger landowners, or potentially wealthy speculators."[6] That was true too of many settlers who arrived from the Upper South, and in their case a "stick" was pushing them as well as a "carrot" attracting them, for land titles were notoriously contestable "at law" in the Upper South states they were leaving behind—Kentucky, Tennessee, North Carolina, and Virginia. Frustration with unreliable land titles prompted many a land-hungry family (including that of Abraham Lincoln) to leave the Upper South and move north across the Ohio River.[7] As of 1850, almost 7 percent of Indiana's residents had been born in Kentucky and another 10.3 percent had been born elsewhere in the Upper South. If the 12.2 percent born in Ohio are added to these (and if the 5.5 percent born in foreign countries are *not* counted—nor of course the 52.8 percent born in Indiana itself) that leaves only about 8 percent who had been born elsewhere in the U.S. Of those, 5.3 percent had been born in either Pennsylvania or New Jersey, and only 3.5 percent had been born in New York State or New England.[8]

Indeed, as of ten years later in 1860, 27 percent of Indiana's household heads had been born in a southern state, which was one percent more than had been born in Indiana itself.[9] Wherever they came from, they settled Indiana by multiplying small, close-knit rural communities, proliferating innumerable local neighborhoods that overlapped each other on all sides. River valleys were usually settled first if they were not marshes. In each locality, the earliest newcomers tended to gain ownership of the best land (unless some land speculator already had) and, later on, their children and subsequent heirs tended to stay put there. Meanwhile, later newcomers often came and went, frequently moving on in search of a place where they too could make good, or at least could do better.[10]

Until the 1830s, rural Indianans' self-sufficient lifeways changed little. It was a laborious way to live, and even after the 1830s it remained laborious for those who were not well-off enough to hire adult helpers or to acquire labor-saving devices. Almost all farm families could reproduce workers in the form of their own children, and they could also benefit from farm animals' reproduction, including the reproduction of draft animals who shared their work in the fields and forests. But not all expansion came thus free of charge. Starting in the 1830s, when only *some* farm families could afford to buy the new labor-saving machines such as planters, cultivators, wheat reapers and hay mowers—and when only *some* could afford to pay a threshing crew to bring a machine and efficiently thresh the harvested wheat—class and social distinctions began growing sharper. Poorer farm families continued performing much of their farm work with hand tools—dibble sticks for planting, scythes and pitchforks for harvesting, bare hands for shelling corn. They were aided by some draft animals, but fewer than the well-to-do families owned. Virtually everyone had at least a plow and one or two draft animals, along with a milk cow or two and some pigs and chickens.

Here's one example of the economic disparities. In 1819 a farmer in far southeastern Indiana who had a surplus of hay inveigled a local handyman to invent a hay-baling device—"an old-fashioned, wooden screw press," as a local historian later called it. Its large wooden screw was turned by horses that were harnessed to the far end of protruding poles and walked in a circle around the haypress. The invention enabled prosperous farmers to grow yet more prosperous by sending baled hay down the Ohio River on flatboats, which proved quite profitable. Many of southeastern Indiana's large farms were soon specializing in hay and their owners were soon among the state's most prosperous farmers.[11] By the mid-1830s, five to ten thousand tons of hay were being shipped each year down the Ohio River from Lawrenceburg in southeastern Indiana.[12] Yet that same southeastern part of the state also contained, and even a hundred years later it still held, poorer farm families who left their hay unpressed (unbaled) and felt lucky if they had even a pulley—that is, a block and tackle rig with which to lift the hay up into the barn loft by using a horse, rather than pitching it up there by hand.[13]

Not all Indiana farmers knew an inventor nearby, but whether they were rich or poor, Indiana's early farm families all depended on access to three kinds of specialists: blacksmiths, millers, and storekeepers. They needed a blacksmith to keep their plows, guns, and other tools in working condition. They needed a miller to grind their corn into cornmeal and their wheat into flour, and perhaps to provide sawed lumber for some of their building purposes. And a storekeeper they needed to trade with—someone who would take their ginseng, eggs, feathers, honey, linen, whiskey, and other home-made products in exchange for things they did not grow or make themselves, like coffee, salt, nails, pots and pans, molasses (at least until sorghum-growing reached Indiana in the 1850s),[14] tobacco (although many grew their own tobacco all along), and some cash they could use to pay taxes and maybe to buy more land.

In the earliest period, storekeepers, millers, and blacksmiths apparently did almost all their business by bartering goods and services in-kind. "Examination of account books," one scholar found, "reveals that transactions for money were comparatively rare and usually received a special notation."[15] If no storekeeper, miller, or blacksmith were nearby, the pioneers would periodically travel great distances to reach one.

Another entrepreneur with whom early Hoosiers did considerable business was the local drover. Each fall, farm families would convey their "surplus"[16] livestock to a drover, preferably for cash but often on a commission basis, and the drover in turn would drive the livestock to a larger drover's gathering point or else drive them directly to a river town such as Cincinnati or Madison where pork-packing took place. All the sizable Ohio River towns acquired pork packers and many Wabash River towns did too.[17] Cattle, on the other hand, often had to be driven further to get a good price for them, as we'll see in Chapters Five and Eight.

Of course, livestock sales were not daily transactions. Pioneers' *daily* dependence was on their farms and on each other. Frequently they swapped work with each other and bartered and borrowed tools and other necessities back and forth. If something they produced was not going to be sold outside the neighborhood, then it was not considered morally right to charge money for it inside the neighborhood. But mentally, if not on paper, almost everyone kept track of who owed them favors and whom they owed favors to. The time-lapse between receiving and repaying a favor was left to mutual convenience. This basic, localized *modus operandi* was an economic system; it could be called a system of "voluntary reciprocity." In Illinois some pioneers called it "the borrowing system."[18]

Thus the pioneers, says one early historian who grew up on a Hoosier farm, "believed and practiced a community of work, but there was an individual score kept. The man who did not help his neighbor roll logs received no help in return, unless on account of charity. . . . They borrowed and loaned with the greatest freedom everything from a team and wagon down to a set of pewter spoons. Yet there was little partnership in the ownership of property. Each family lived to itself and had no great desire to have near neighbors."[19] (Anyway most of the *men* seemed to have little desire for near neighbors. How women felt we'll get to in Chapter Six below.)

At first almost everyone farmed, even doctors and lawyers, preachers and teachers, not to mention storekeepers, blacksmiths, millers, tanners, and drovers. If a neighborhood was not heavily settled yet, land would be available to use (although not necessarily affordable to own) by almost anyone who was willing and able to farm. Non-landowners generally had to sharecrop. Few of them could afford to pay rent in cash to a landlord for their farm tenancy. In the early days, the sharecropper's share of the crops was usually more than the landowner's share, but that ratio shifted as settlement grew thicker and thus landowners could bargain to get more of the crop.[20] Many of the non-landowners, even if they were already operating a farm on someone else's land, also hired out some of their own or their children's labor to neighboring landowners—helping their better-off neighbors to plant, cultivate, and harvest, and receiving compensation either in cash or as a share of *that* crop. If such a helper brought along a team of draft animals that aided the work, he would receive double compensation.

All of this grassroots economic "contracting" overlapped with all the "voluntary reciprocity" that concurrently was going on, and they could both flourish with almost no cash changing hands—unless someone was trying to save up money to move away or to buy land. Both moving and buying land did usually require an accumulated nest-egg of money.[21]

Farming was hard work, but it wasn't the only source of sustenance. Plenteous food in the form of wild animals, fruits, and herbs was available simply for the taking. Even if a local magnate or an absentee speculator held title to a tract of land, if that land was unattended it was used as a neighborhood commons for hunting and gathering and also for livestock foraging. Pigs and cattle were allowed to run loose on such land to forage for themselves, with their ears notched to identify their owners.[22] Thus, the fields that were used for growing crops had to be fenced in, which was usually done with Virginia rail fences—also called "worm fences," since they wormed their way across the land. Those fences used a lot of wood, but of course wood was free—at first, anyway.

Lest this picture sound idyllic, let's look at some of its downside. It was narrow and confining, especially for women and girls since they were expected to stay home and do gender-defined chores. The price of self-sufficiency with *respectability* was continuous toil for womenfolk, and "slacking off" brought lowered status in the neighborhood and greater dependence on other families.[23]

Meanwhile, many pioneers experienced an intense religious life that was fed by troubling uncertainties about their immediate personal futures. Native Americans' intentions, weather conditions, crop prospects, livestock ailments, personal accidents, fevers, and a host of other worries fostered a sense of dependence on God's beneficence. Religious meetings tended to be emotional.[24] Rural churches were generally led by successful male farmers as elders. Preachers would visit "on circuit."

All the while, many Hoosier pioneers were working so hard that they shortened their lifespans. The diligence of their drive for "improvement" frankly startled some European observers,[25] but it was not mindless. The Midwest's 1800s "improvement" craze grew directly out of the new dispensation that had been proclaimed in the United States Declaration of Independence and then consolidated by the Constitution, and reinforced yet more by the government virtually giving away the fertile Midwest in spacious parcels of land that thereby became private property.[26] One historian says, "put simply, free property made free men and women."[27]

And that is why—to flip back to where we left Thomas Jefferson—Indiana's settlers did *not* really feel they owed "a decent respect to the opinions of mankind." Instead, they thought they "had it made." As an editor in Elkhart proclaimed in 1859, "We are not headed for another crash . . . but for the heyday of prosperity and money making."[28] Admittedly the Civil War put a painful parenthesis in their self-confidence, but that proved transitory. After the Civil War, material aims intensified as the supply of enticing economic niches grew scarcer and Hoosiers grew more competitive toward each other. After the Civil War, fewer profits could be gained through pure speculation—by just buying land, for instance, and "waiting for a rise."[29] That benign path to many pioneers' success had started drying up by the mid-1800s. Instead, to gain profits by investing money came to require employing other people, such as by hiring them to perform labor on crops, or with livestock, or in mines or factories. The middle third of the 1800s saw sharper class lines emerge in Indiana, the kind of trend that some historians have called "Europeanization."[30]

Thus, the years after the Civil War saw the "free labor" ideal of Abraham Lincoln and other Republicans start looking a bit threadbare. In his December 1861 State of the Union address, Lincoln had asserted that

> many independent men everywhere in these States, a few years back in their lives, were hired laborers. The prudent, penniless beginner in the world labors for wages awhile, saves a surplus with which to buy tools or land for himself, then labors on his own account another while, and at length hires another new beginner to help him. This is the just and generous and prosperous system which opens the way to all—gives hope to all, and consequent energy and progress and improvement to all.[31]

Small-scale enterprises did still enable many people to fulfill Lincoln's economic ideal, but small-scale enterprises were growing more difficult to sustain. Family farming, for instance, did still offer economic independence for many of the families engaged in it. But in Indiana's earliest-settled parts, farming was already losing its "easy entry" traits by the 1830s, and in most of the rest of Indiana that was happening by the 1850s. The era when anyone and everyone could be a farmer was ending, and quite quickly.

In Chapter Eleven we'll look at the new class distinctions that then emerged, distinctions derived from wealth rather than ones conferred by birth. Jeffersonians were still numerous and vigilant, and they fought back against those class distinctions through a movement called "producerism," one of whose spokesmen in fact was Abraham Lincoln. In his 1861 State of the Union speech quoted above, Lincoln advocated an America in which "men with their families—wives, sons, and daughters—work for themselves, on their farms, in their houses, and in their shops taking the whole product to themselves." And *why* should the whole product go to its producer? Because, said Lincoln, "Labor is prior to, and independent of, capital. Capital is only the fruit of labor, and could never have existed if labor had not first existed. Labor is the superior of capital, and deserves much the higher consideration."[32]

As a widespread aspiration for American society, this goes back at least to Jefferson and the Democratic-Republican Party that he and others started in the 1790s. But they had been so intent on overthrowing "ascribed" status (conferred by high birth) that they failed to guard vigilantly against the possibility of inequality based on wealth. Jefferson visualized easy-entry farming protecting American equality "to the thousandth generation." In 1805, while president, he wrote that American wage workers were "as independent and moral as our agricultural inhabitants" and that "they will continue so as long as there are vacant lands for them to resort to; because whenever it shall be attempted by the other classes to reduce them to the minimum of subsistence, they will quit their trades and go to labouring the earth."[33] But note Jefferson's precondition: "as long as there are vacant lands for them to resort to."

In fairness I should add that Jefferson as early as 1785 had started advocating measures to guarantee that poor people *had* access to land. He wrote: "the earth is given as a common stock for man to labor and live on. If for the encouragement of industry we allow it to be appropriated, we must take care that employment be provided to those excluded from the appropriation. If we do not, the fundamental right to labor the earth returns to the unemployed [and] it is not too soon to provide by every possible means that as few as possible shall be without a little portion of land."[34]

In that same 1785 letter, Jefferson also urged that smallholders should be "exempt . . . from taxation below a certain point" and that "higher portions of property" should be taxed in "geometrical progression as they rise."[35]

The old easy-entry farming of the pioneer era required *three* preconditions, actually. Jefferson's idea of making landownership easily available was only one of them. All three were: (1) that the needed skills (as assigned by gender) be taught free of charge to rural children, (2) that the needed land be easily available (which was Jefferson's point), and (3) that the needed tools as well be easily available.

Of these three preconditions, the first stayed pervasively true for another one hundred years and more, for farm children did indeed learn the trade. In many neighborhoods they were still learning it after World War II.[36] But the other two preconditions (easily available land and tools) ended in Indiana in the mid-1800s. The third precondition, that of cheap tools, saw a gradual rise in the cost of farm implements start in the 1840s toward what would eventually in the mid-1900s become enormous outlays, financed generally on credit. In Chapter Eleven we'll glance at that and also at what happened to land prices in the mid-1800s as a profusion of railroads crisscrossed Indiana.

The upshot was that more and more people who would have liked to be independent farmers did not achieve that dream. As of 1878 the semi-monthly *Indiana Farmer* was still editorializing in Lincoln's vein, asserting that most successful farmers had begun as "farm laborers, and by industry, thoughtfulness, and economy have worked their way into fair possessions and easier circumstances." But the same 1878 issue of the *Indiana Farmer* also carried a second editorial reporting that unemployed farm workers who were "anxious to work" that summer were instead experiencing destitution and even starvation.[37] Unemployed farm hands that year in Indiana were offering their labor for $10 a month without board. They just wanted enough wages to feed themselves. Some were so desperate that they went around destroying reapers and binders so that farmers would need their services. A few of them burned down a farm-implement factory in east-central Indiana's town of Liberty. Indiana farmers were often sympathetic, and many of them "set aside their machinery and hired workers to cut wheat by hand" with cradle scythes, but they weren't willing to pay much for the work.[38]

And there, in the late 1800s, rather than launch into the twentieth century with all its complicated new issues, we will try to plant ourselves and take stock of why Indiana's farm families had already so quickly passed from pioneering to persevering.

Chapter One

Native American Agriculture before European Contact

> Anyone who has lived with primitive tribes, who has shared their joys and sorrows, their privations and their luxuries, who sees in them not solely subjects of study to be examined like a cell under the microscope, but feeling and thinking human beings, will agree that there is no such thing as a "primitive mind," a "magical" or "prelogical" way of thinking, but that each individual in "primitive" society is a man, a woman, a child of the same kind, of the same way of thinking, feeling and acting as a man, woman or child in our own society.
>
> —Franz Boas, *Primitive Art*, p. 2

Farming can be viewed from many angles. If we consider not only financial returns but also long-term sustainability—and if we consider all the human values that are linked to family farms—then we can find much to learn from early Native Americans' way of farming. What kinds of expertise went into Native Americans' development of wonder crops such as squash, corn (maize), potatoes and beans isn't all understood. But it *is* understood by now that Native Americans grew those crops in ways that preserved the soil and were able thereby to continue indefinitely. They chose bottomlands where trees were relatively small. Men and women worked together, girdling the trees and digging out the brush to burn it. "After a time—five or ten years, or perhaps twice in a generation—the Indians customarily abandoned their clearances and commenced new ones."[1]

Europeans reached what is now Indiana in the persons of the Frenchman René-Robert Cavelier, sieur de La Salle and a bevy of companions in December 1679. By that time, Native Americans in the Midwest had been cultivating plants for about 5,750 years, beginning with a gourd-like "pepo squash" and also with sunflowers, goosefoot (lamb's-quarter), and marsh elder (sumpweed). They had also been cultivating corn here (or nearby) for approximately 1,800 years, and beans for about 400 years.[2]

One reason why Native Americans didn't use land-damaging farming methods is because they domesticated no draft animals. Prehistoric North American Indians domesticated dogs and turkeys, to which South American Indians added only llamas, alpacas, and guinea pigs. Mesoamerican Indians added a few small fowl.[3] The Indians' lack of draft animals kept their agriculture essentially horticultural and labor-intensive.

If we wonder why Native Americans didn't domesticate draft animals and train them to pull plows, we can start with the fact that no likely animal candidates were present in North America—no large cattle or horses. Even after Europeans came, many Indians were reluctant to use draft animals. As late as the latter 1700s, for instance, after one hundred years of Cherokee-European interaction, many Cherokees remained unwilling to use draft animals and plows because that would mean "those farmers who used the hoe and the dibble stick would not be able to compete, and the result would be unemployment and starvation for the elderly."[4]

Beyond Indians' concern for their own communities, Indian agriculture was enmeshed in the idea that all forms of life participate in a great community of reciprocal gift giving. Through rituals and shamanism, Indians sought to sustain awareness of their community's reciprocal relationships with various animals and plants.[5]

Beliefs and rituals of non-literate people leave no obvious physical remains, so the spiritual aspects of prehistoric agriculture would be hard to reconstruct. Indians did, however, sometimes bury plants with the bodies of their deceased. Plant remains have also sometimes been found in ritual structures.[6]

Many of the world's tradition-minded farmers still perform rituals at planting time, and still hold festivals at harvest time, and it is known that Native Americans did so in what is now Indiana. In early August of 1791, a member of a United States military expedition under General James Wilkinson reported harvest festivities at the Miami people's town of Kenapacomaqua, which lay on the Eel River near its confluence with the Upper Wabash River:

> The surprize of this town was so very complete, that before we received orders to cross the river and rush upon the town, we observed several children playing on the tops of the houses, and could distinguish the hilarity and merriment that seemed to crown the festivity of the villagers, for it was the season of the green corn dance.[7]

Besides killing eleven warriors, capturing forty women and children, and burning all the houses, the American troops succeeded in "destroying about 200 acres of corn; which was then in the milk, and in that stage when the Indians prepare it for Zossomanony. This success was atchieved with the loss of two men, who were killed."[8]

Today, many people wish to better understand what they share with humans of diverse cultures across chasms of time and differing world views. Some mental and behavioral patterns are shared by all humans; others are culturally specific, acquiring their significance within a single culture's unique web of interconnected meanings. In prehistoric times, not only hunting and gathering but also agriculture involved far more intimacy between humans and other forms of life than those activities usually involve today.[9]

The breaking away from such intense human-animal-plant relations began, plausibly, when humans domesticated various species of herbivorous animals and established herding-based societies.[10] As villages became permanent sites and some became cities, one form of power that their leaders claimed was spiritual power. New leaders evidently claimed that the most important spirits dwelt not in other beings but in city temples or in the heavens—which plausibly (perhaps as an afterthought?) helped people feel free to simply "use" nature, including other beings, without regard for mutuality.

By contrast, people in traditional societies which lack herding tend to seek mutuality with animals and plants. It is customary to become a full member of such a society through intense personal experiences with nature. Many Native American societies encouraged their young people to embark on a vision quest, seeking an especially personal relationship with one particular type of animal or plant.[11] This can be thought of as a form of acculturation (an acculturation with nature), and it formed the basis of teenage initiation in many tradition-oriented societies. It was meant to create a relation of mutuality between young people and other forms of life and was regarded as the gateway to full membership in society.

One way to interpret such initiations might be to hypothesize that everything alive possesses two foci of existence simultaneously, one exterior and the other interior. The interior existence of each type of being can then be thought of as unified and also as self-conscious. In other words, even though deer, beaver, bear, and corn possessed no individual self-consciousness, nonetheless self-consciousness might have occurred for a Deer Spirit, a Beaver Spirit, a Bear Spirit, a Corn Spirit. Each of those spirits might be (or might once have been) not only self-conscious but able to communicate. And humans might have been able as separate individuals to communicate with Deer Spirit, Beaver Spirit, Bear Spirit, Corn Spirit. Among Algonkians, such spirits of various kinds of animals were called "keepers of the game," and Algonkian shamans would negotiate with them issues such as the number of animals which hunters could legitimately kill to sustain their people.[12]

Communication for the sake of mutuality with other types of life may sound exotic, but in some parts of the world it apparently still goes on. What we call a civilized environment is what some of our human contemporaries call "the wilderness."[13] People whose sense of order is interior tend to experience external criteria of order as disorder.

More prosaically, hunter-gatherers experience less anxiety about food supplies than do members of fully farming-dependent societies.[14] Within the last few centuries of "Mississippian culture" prior to European contact, Native Americans in what is now Indiana devised a lifestyle that entailed considerable agriculture during the growing season, but even during that warmer part of the year they continued to hunt and gather, and after harvesting and drying their crops they then spent the cold season in small nomadic hunting bands.[15] Thus they were generally able to acquire food and other necessities in various ways. If one means of sustenance temporarily failed, they generally could turn to another means—which often is not true for full-time farmers.

Archeologists now estimate that humans have been in the Western Hemisphere for up to 30,000 years. On the basis of surface finds and excavations, Indiana archaeologists distinguish four eras in Indiana's prehistory. The recession of the "Wisconsin" glacial ice sheet permitted a Paleo-Indian era from about 9500 B.C. to about 8000 B.C. In that era the primary game animal was the mastodon, hunted by Paleo-Indians whose spears held stone points that were "fluted." (That is, down the middle of both flatter sides, their stone spear points were grooved to allow blood to flow freely from wounded mastodons.)

As the mastodons suffered extinction, the hunters turned primarily to a large species of bison, and hunters no longer fluted their spear points. This inaugurated the Archaic era (8000–1000 B.C.). After several millennia, at approximately 5000 B.C., the large bison followed the mastodon into extinction, and thereafter people gave less emphasis to big-game hunting than they gave to a flexible combination of small-game hunting, fishing, and gathering.[16]

About two thousand years before the earliest known midwestern agriculture, several locations in the Midwest already held people who lived a sedentary lifestyle. Proto-urbanization thus likely began in the Mississippi Valley long before people domesticated plants there. Those proto-urban people subsisted largely on mussels (shellfish), fish, hickory nuts, and deer meat at settlements such as the Koster site on the Lower Illinois River.[17]

No clear line of demarcation separates the "gathering" of wild plants from the "cultivation" of domesticated plants. Women were gatherers whereas men were hunters. While foraging, prehistoric women might manipulate a plant in ways that caused the plant to produce more of what people ate. Such manipulation might be intentional or *un*intentional. Just disturbing the ground might lead to more food production. Loosening the soil could encourage Jerusalem artichokes and other wild tubers to spread. Similarly, burning sections of land might enhance not only the growth of grasses attractive to game animals but also the growth of wild foods that people ate. Furthermore, wild food-bearing plants might be tended by weeding around them, watering them, or pruning them. And they could be transplanted, which could lead to creating a garden—all without yet sowing seeds or saving seeds from year to year.

Or, by contrast, the first gardens might have simply grown unintentionally at refuse dumps near campsites where seeds of wild edibles were thrown away. Whatever the case, with time savvy gardeners were saving seeds from plants with the traits that they most desired, including plants that were ready the earliest for harvest, thereby selecting for earliness and so on for other desired traits. The definition of a fully domesticated plant is one that's so modified by human intervention that its survival depends on continued human intervention. Today's mass-produced corn and wheat are two examples.[18]

Evidence now suggests that by 3000 B.C. some of the people who then lived in present-day west-central Missouri were cultivating a gourd-like "pepo squash" domesticated either there or elsewhere in the midwestern region. The pepo is actually a form of fruit, an extra large berry almost the size of a tennis ball with a heavy, leathery rind that resembles a squash or gourd rind. As for *true* squash, the earliest of the Midwest's known sites where true squash was grown is at Phillips Spring in west-central Missouri.[19]

Today's midwestern United States is one of only seven areas of the world where there exists no doubt that plants were domesticated from the wild. Approximately concurrent with the domestication of "pepo squash" in the Midwest, several other plants were also domesticated there—sunflower, goosefoot (lamb's-quarter), and marsh elder (sumpweed).[20]

All this far predated the era known as Eastern Woodland. The Early and Middle Woodland era is dated from about 1000 B.C. to A.D. 900. (Another name for the Early Woodland period is Adena, and the Middle Woodland period has often been called Hopewell.) By 500 B.C. at the latest, domesticated plants were providing a significant portion of midwestern natives' diets. And by about then, several additional plants had been domesticated in the Midwest—knotweed, maygrass, and little barley. Corn (maize) entered this mix soon afterwards, coming as an import from Mesoamerica via the Southwest, but at first corn was not much grown. Indeed, it was virtually ignored while the seeds of sunflower and marsh elder changed dramatically between 200 B.C. and A.D. 400, indicating that purposeful selection was influencing those two plants.

Corn finally was widely (but still sparsely) cultivated in what's now the eastern United States about A.D. 800. Then suddenly by A.D. 1000 corn was being *heavily* grown. Corn-growing expanded through the use of northern flint corn that held twelve rows of kernels on its cob. Northern flint is a variety with a relatively short growing season. Later, Indians introduced it to Euro-American settlers, and later still, when it was cross-pollinated in the Midwest with "gourdseed" varieties from the South, the result was the "dent" varieties that pervaded the corn belt and led to today's ubiquitous dent hybrids.[21]

Extensive moundbuilding, incidentally, had begun early during the Woodland period, and moundbuilding continued almost until the time of European contact. Some scholars think that the small amounts of corn grown before about A.D. 800 were grown specifically for use in ceremonies conducted at mounds. When Spaniards first reached the lower Midwest, such as during De Soto's 1539–1542 expedition, they found that "noble allies" of "paramount rulers" were tilling sacred corn fields that were kept separate from other corn fields.[22]

For Indiana purposes, the era between A.D. 900 and European contact is called the Middle Mississippian/Late Woodland/Upper Mississippian era. Recent studies of Native American bone collagen in the "Eastern Woodlands" territory suggest that a typical Native American diet as of A.D. 1000 consisted 24 percent of corn. But then by A.D. 1200 the diet consisted 45 percent of corn, and by A.D. 1300 consisted 70 percent of corn.[23] And that's when the cultivation of beans apparently reached the Midwest—around A.D. 1300, which was two or three hundred years later than bean cultivation had reached present-day New York State.[24] Like corn, beans had originated far to the south.[25] Eating beans and corn together, as in the Native American dish called succotash, greatly enhanced their separate nutritional benefit. The combined amino acids of beans and corn provide virtually everything the human body needs to create human protein.[26]

Also by this time (A.D. 1300), tobacco and perhaps several other plants were being cultivated in what is now Indiana. And it seems likely that wild rice was being systematically harvested from marshes and lakeshores in what's now northwestern Indiana, which remains today a marshy part of the state wherever it is left undrained. The Miami Indians were not known for gathering wild rice but the Potawatomi were.[27]

Just as women had been the gatherers in the hunting-and-gathering lifestyle, likewise women became the farmers and gardeners in horticultural days.[28] The usual size of cornfields during Woodland and Mississippian times seems to have been from one-third of an acre to one full acre. Squash planted alone probably produced three to four tons per acre (before the squash was dried). But only about half that much squash could be expected if squash was interplanted in the same field with corn. Yields of corn probably averaged around twenty-five bushels of shelled corn per acre, enough to nowadays fill a pick-up truck, and that yield would not have been diminished by interplanting with squash. As for beans—this far north most of the beans would have been pole beans (rather than the more southerly bush beans) and thus they would have benefited from interplanting with corn, the cornstalks serving as their poles.

Cultivation in prehistoric Indiana was often perhaps carried out after slash-and-burn land clearing. Van A. Reidhead estimates that in what's now Indiana, crops were grown on each plot of prepared ground for only about three years. Then followed, he says, a long fallow period lasting at least two or three decades.[29]

Slash-and-burn methods are still commonly used by many farmers in the tropics. They need not be environmentally destructive when used by a sparse population who limit their burnings to small fields. Such was the situation in prehistoric times but it is often not true today where slash-and-burn agriculture remains in use.[30]

Using stone and wooden tools, the number of work hours required to prepare land for planting has been estimated at 415 hours per acre—thereby averaging 138 hours per year for the preparation of a field that, using Reidhead's estimate, would stay usable for three years.[31] The brush would be cut down and the large trees would be girdled. After both the brush and the deadened trees had dried awhile, the brush would be burned at the base of the trees. Then, using hoes made of wood, shell, stone, or animal shoulder bone, the ashes would be worked into the soil in the process of heaping up hillocks of soil to receive seeds. Corn and bean seeds would be planted together in many of the hillocks, and squash seeds would be planted in the rest of the interspersed hillocks—all under the withered leaves of the dying trees.

Seed-planting was aided by the use of dibble sticks for poking holes and it required only about ten work hours per acre for each of the three crops planted—corn, beans, and squash.

When corn *was* indeed planted, the task of protecting the crop required an estimated 167 work hours each year per acre. More birds and other animals raided ripening corn and beans than raided ripening squash. The task of protecting crops could be performed by children.

An estimated annual average of sixty-seven work hours went into weeding. During a field's first year, not much weeding was needed, but after a field's third year, Reidhead thinks so much grass and so many weeds would be intruding, and the soil's fertility would by then be so depleted, that the farmers were likely to abandon that field and begin clearing brush off a fresh piece of land.

Harvesting required about seventy hours per acre for corn and twenty hours for the interspersed squash. No estimate is available on how long an acre of cornstalk-climbing beans took to pick. (Having personally picked both beans and corn, I'd guess it took a lot longer to pick an acre of beans than an acre of corn. But had I been an Indian mother who looked forward to feeding those beans to her children all winter, perhaps my own bean-picking days wouldn't have *seemed* so long.)

Constructing storage facilities demanded about forty hours for the corn that was harvested and another forty hours for the squash harvested. Pulling back the shucks of the ears of corn, then tying the ears together and hanging them inside to dry would also have taken about forty hours—whereas slicing and drying the squash would have taken much longer, about 127 hours.

Adding all of this together, and also adding 261 estimated work hours for harvesting, drying, and storing the beans, an investment of about 1,000 work hours per acre would have been required in order to provision a group of Native Americans in prehistoric Indiana with twenty-five bushels of corn, 3,000 to 4,000 pounds of squash (before weight loss through drying), and an undetermined supply of beans.[32]

As mentioned, the size of prehistoric fields ranged generally from one-third to one acre. Native Americans also commonly planted much smaller gardens. But apparently rare prior to European contact were large fields—except at a few locations, and in Indiana apparently only at the Mississippian "Angel" site which spread across a broad terrace above the Ohio River near the location of present-day Evansville. Between approximately A.D. 1200 and 1600 this site was home to about 1,000 moundbuilders. Several smaller Mississippian (and/or Late Woodland) moundbuilder sites likewise existed along the Ohio, Wabash, and White rivers in present-day Indiana.[33]

Extensive moundbuilding had begun early in Woodland times and it continued almost to the time of direct European contact. Prior to direct contact, European diseases were already decimating the populations of Great Lakes-area Indians, and those diseases might have been a factor in dispersing the populations of moundbuilders. Written records of epidemics did not begin, of course, until after direct European contact.[34]

CHAPTER TWO

Native American Agriculture after European Contact

> . . . [I]T WAS A BEAUTIFUL PLACE, two or three miles below the mouth of the Tippecanoe River. . . . The waters of the Wabash were rich in fish, and turtles deposited their eggs on the islands and sandbars that abounded. Narrow tablelands, covered in apple, maple, sycamore, and wildflowers, and dissected by springs that coursed down the hillsides, bordered the Wabash, and there, as well as in the hills, prairies, and groves of trees behind, lived a profusion of wildlife. . . . Running five hundred yards below the town were the cultivated fields, more than one hundred acres.
>
> —John Sugden, *Tecumseh*, p. 167

The future state's first documented visit by a European occurred in December 1679 when René-Robert Cavelier, sieur de La Salle and a bevy of fellow *voyageurs* traveled down the eastern shore of Lake Michigan, stopped at the mouth of St. Joseph of the Lake River, built a small fort there, journeyed up that river, portaged over to the headwaters of the Kankakee River near the present-day location of South Bend, and journeyed down the Kankakee to the Illinois River.[1]

La Salle was a fur trader. Arrangements followed for Indians to deliver furs to he and his agents, and in exchange to receive European metal goods and other trade items. These early exchanges centered around a fort and trading post that La Salle established in 1682 at Starved Rock on the Illinois River. Thousands of Indians resided in that vicinity through the mid-1680s, and thousands more resided east of Starved Rock along the Iroquois River near the present boundary between Illinois and Indiana.[2]

Prominent among the manufactured goods which Indians then acquired were iron tools and weapons that drastically reduced the amount of work in their daily lives.[3] Viewed from today, with today's hindsight into the last 325 years, the European invasion appears first and foremost as a *threat* to Native Americans, but that aspect of Europeans' presence was not always uppermost in Indians' actual experience. Indeed, the Europeans must initially have seemed peaceful compared to the Iroquois, whose massive attacks on midwestern Indians had begun in 1648 and had virtually emptied Indiana of its previous occupants.[4] By the late 1680s, however, the Iroquois threat had subsided, and by the early 1700s Indiana was being repopulated by Miami and other peoples.[5]

Iron products (and also some of bronze) were supplied by La Salle and his agents and they led to major changes in Indians' lives. Here some conjecture is necessary, but when twentieth-century *steel* tools are substituted for stone, wood, bone, or shell tools used in slash-and-burn farming, the work hours needed to prepare an acre of ground for planting plunges from about 415 hours to about 69 hours. The work hours needed for weeding also decline somewhat.[6] The significance for one tribe in the Amazon Basin can be gauged from an anthropologist's report that "the highly productive food economy of the Machiquenga depends on metal objects obtained from Peruvian traders. Without an outside source of axes, the Machiguenga would have to give up their semisedentary existence and roam the forest as nomads."[7]

Plausibly, iron tools supplied to Indians by Europeans in the Midwest helped to facilitate the creation of large corn fields such as were observed in Indiana after Miami, Potawatomi, and other peoples began congregating there in the early 1700s.[8] A French trading post was established near the Wabash River at Fort des Miamis (close to today's site of Fort Wayne) by the year 1697, another was established at Ouiatenon (close to present-day Lafayette) in 1719 or 1720, and a third was started at Vincennes by 1731.[9] One observer in 1718 reported seeing near Ouiatenon fields of the Wea people stretching for a length of six miles and containing "Indian corn, pumpkins and melons."[10]

Later, the member of the August 1791 American military expedition under General James Wilkinson whose report we glanced at back in Chapter One on page 13, reported that that expedition destroyed 200 acres of corn "in the milk" at the Miami village of Kenapacomaqua on the Eel River near the Wabash, and then destroyed another 200 acres of corn at Kickapoo villages to the west.[11] Then in 1794, the year of the Battle of Fallen Timbers, General Anthony Wayne wrote of the battlefield vicinity that "the margins of these beautiful rivers, the Miamis of the Lake [i.e., the Maumee River], and the Au Glaize [a southern tributary of the Maumee], appear like one continued village for a number of miles, . . . nor have I ever before beheld such immense fields of corn in any part of America, from Canada to Florida."[12]

After summer 1794's Battle of Fallen Timbers, United States troops spent many days destroying such cornfields. A member of the expedition wrote that cornfields extended four or five miles along the Au Glaize, encompassing a thousand acres of growing corn. He added that the whole valley of the Maumee from its mouth on Lake Erie west to the site of Fort Wayne was full of immense cornfields, large vegetable patches, and old apple trees.[13]

Since it was an array of iron goods acquired from European traders that apparently made possible such prolific Indian agriculture, why didn't Indians start making iron on their own? Indeed, some *copper*-making had occurred prior to Europeans' arrival. From about 1200 to 700 B.C. an "Old Copper Culture" had flourished in northern Wisconsin and adjacent areas. Copper spear-points, knives, and even axe-heads have been found at Old Copper Culture sites. Indiana itself contains little copper ore, and few prehistoric copper fragments have been found in Indiana, but evidence exists that copper was also smelted in places other than Wisconsin, and at other times.[14] What's perhaps crucial, however, is that copper's softness makes its tools far less useful than iron tools. To liquidify copper requires a temperature of only 1,981 degrees Fahrenheit whereas to liquidify iron takes 2,795 degrees. No sign has been found of Native Americans having smelted iron, either before *or* after European contact.[15]

By contrast, iron-making often became one of the very first industries started by the Indians' nemesis—namely, by the Euro-American settlers who began pouring into the Old Northwest in the 1790s. That left Indians more dependent on trade than were those newcomers, many of whom also made their own guns and gunpowder.[16]

The well-known agricultural historian R. Douglas Hurt asks why Native Americans, "who already had a long tradition of farming and who were among the most skilled" farmers anywhere in the world, were not considered by most Euro-Americans to be successful farmers. Hurt asks, "What were the environmental and cultural limitations that prevented Indian farmers from emulating white farmers?"[17] In hindsight we might be thankful that Indians farmed with limits, but Hurt's question also brings to mind something that Tecumseh and his brother the Prophet pondered—the economic and then political dependence that enmeshed Indians who did things the Euro-American way. And another scholar points out that "a large part of the reason for the discontinuation of the Native American food systems in the Midwest (and elsewhere) was that European-Americans were able to successfully marshal the resources to stake and defend claims to the land rather than any inherent failure of the former system."[18] Indians lost control over Indiana because they were outnumbered. The 1790 federal census found 73,677 Euro-Americans already living in Kentucky alone, which may have exceeded the number of Native Americans in the entire Ohio River watershed.[19]

Perhaps we can also better understand what happened by contrasting what Euro-Americans and Indians generally wanted from each other. What Euro-Americans generally wanted from Indians can be summarized in four categories: (1) Some Euro-Americans wanted Indians to take part in enterprises in which capital could be invested and profits could be earned. (2) Some wanted Indians to participate as consumers in the market for European or American manufactures. (3) Some wanted Indians' acquiescence in European or American control over natural resources, including land for farming. And (4) so as to help Euro-Americans achieve such access to trade and consumers and natural resources, some Euro-Americans wanted Indian participation in their networks of military or paramilitary activity.[20]

If we look as well at what Native Americans wanted, we can see that Euro-Americans represented more to Native Americans than merely a threatening intrusion. Metal tools for farming, metal knives, guns, cooking kettles and many other conveniences were attractive to most Indians. They could acquire them by exchanging some of their surplus crops with Euro-American explorers and merchants,[21] and particularly by exchanging animal furs and skins. At first, furs and skins were abundantly available to Indians. Cleaning furs and delivering them to trading posts seemed, at first, a small cost to pay for metal objects that facilitated daily tasks.[22]

Despite most Native Americans' embrace of new conveniences—including not just metal objects but cloth yard goods—the idea of unlimited material accumulation was alien to Indian lifeways, which emphasized sharing. By sharing their surplus belongings, people gained prestige and goodwill and could in effect accumulate a "bank account" of potential return favors. Such social credits could also be acquired by providing generous hospitality.[23]

So if we look at Native Americans' desires in tandem with the four Euro-American desires that were summarized above, we can see why most Indians were willing consumers of certain manufactured goods (the Euro-American desire number 2) but we can also see that the other three types of Euro-American desires encroached on Indians' way of life. Yet, in order to acquire manufactured goods, Indians realized they had to share in business activities (Euro-American desire number 1)—especially in the fur trade, a venture in which the Indians' role (as laborers) grew increasingly disadvantageous.[24] And since few Indians became financial accumulators, Indians were rarely business investors. If some business profits did fall their way, those were not usually accumulated or invested financially.

As for desire number 3, Euro-Americans' wish to control natural resources, this eventually administered the *coup de grâce* to Indians' traditional lifeways in Indiana.[25] When thousands of Euro-American farmers demanded control over land for their own sustenance and for potential profits, at that point the very practices that had long kept Indian food sources sustainable suddenly became Indian *dis*advantages. For example, Indians' land tenure was controlled by groups, not by individuals. Indian groups could severely sanction any individual Indian who behaved antisocially.[26] But now Indians faced intruders who guaranteed each other their individual private ownership of land tracts no matter how antisocially some of them behaved.

Encroachment on Indian lands by settlers occurred not so much from New France as from the British colonies. In the 1740s several hundred English fur traders, mainly from Pennsylvania, began operating in the Upper Ohio River Valley.[27] Then on 12 July 1749 the colony of Virginia, with England's permission, granted 800,000 trans-Appalachian acres to the Loyal Company and, on the same day, granted 200,000 acres to the Ohio Company of Virginia. In spring 1750 the Loyal Company sent Dr. Thomas Walker to look for valuable land in Kentucky, and in the autumn of that year the Ohio Company sent Christopher Gist to Ohio to look for "a large quantity of good, level Land, such as you think will suit the Company."[28]

Among the many land scouts and "long hunters" who followed in the wake of Walker and Gist was Daniel Boone. These land scouts often found that the best land for farming consisted of old Indian fields located on bottomlands and terraces adjacent to rivers. In January 1751, adjoining a Delaware village beside the Scioto River in central Ohio, Christopher Gist noted a "plain or clear Field about 5M[iles] in Length . . . and 2 M[iles] broad," and he noticed many other choice sites in the same vicinity.[29]

More than thirty years later, in the 1780s, Virginia turned over most of its western lands to the United States government. But in the process, Virginia took special pains to keep possession of fertile Indian fields along the Scioto River like those that Christopher Gist had extolled during his 1751 visit. They became part of the Virginia Military Tract in Ohio.[30] Similarly in Indiana a few decades later, white settlers often placed their homes and crops precisely where the homes and crops of Indians had previously stood.[31]

Theodore Roosevelt claimed in his 1889 book *The Winning of the West* that the Indians' key disadvantage in their relations with whites was their lack of a land-ownership system comparable to the European system. "Had we refrained from encroaching on the Indians' lands," wrote Roosevelt:

> The war would have come nevertheless, for then the Indians themselves would have encroached on ours. . . . The question which lay at the root of our difficulties was that of the occupation of the land itself, and to this there could be no solution save war. The Indians had no ownership of the land in the way in which we understand the term. The tribes lived far apart; each had for its hunting-ground all the territory from which it was not barred by rivals. Each looked with jealousy upon all interlopers, but each was prompt to act as an interloper when occasion offered. . . . The white settler has merely moved into an uninhabited waste; he does not feel that he is committing a wrong, for he knows perfectly well that the land is really owned by no one . . .[32]

Theodore Roosevelt seemed to imply that European-type landownership could have saved the Indians from expropriation. But is that actually true? Spanish, French, English, and American intruders all claimed that their discoveries and/or conquests in America had conferred legal authority not simply on their own *type* of system but literally under their own jurisdiction. The national United States government did tend to acknowledge Indian land claims that other Indians acknowledged—but at most to merely the extent of acknowledging legal ownership, never to the extent of acknowledging legal jurisdiction.[33] Indeed, during the first fifty years of American independence, Indians ceased to be considered landowners and came to be viewed as mere occupants of the land.[34]

Then too, United States authorities sometimes applied their laws elastically to achieve politically expedient results. In Indiana Territory in 1812, for instance, "when war was officially declared, initial expeditions focused on areas of resistance to [William Henry] Harrison's treaty progress, such as the Miami villages along the Lower Mississinewa. During the course of the war at least twenty-five Indian villages and towns in Indiana were struck and destroyed. These included villages (and crops) of Miami, Potawatomi, Kickapoo, Winnebago, Delaware, Nanticoke, and Wyandot. Of the nineteen total campaigns in Indiana, fifteen were initiated by United States troops or militia."[35]

On 3 May 1814, in another instance of expediency, a United States official wrote to another official:

> With the view of increasing [the Indians'] wants and distresses and thereby rendering them more harmless, I have permitted all the traders to sell as much liquor as they thought proper. This in a political point of view, at this time, is of more effect than many would suppose.[36]

The treaty toward which this policy aimed was concluded at Greenville in northwestern Ohio in July 1814. Its purpose was to convince neutral Indians (including the Miami) to attack the British. Liquor flowed freely to Indians during the negotiations.[37] Some of these same Indians, including Miamis, were then coerced in 1818 to sign the New Purchase, which allegedly extinguished all Indian claims in the new state's central one-third. Among those ousted by the New Purchase were approximately 1,000 Delaware, Muncees, Moheakunnuks, and Naticokes who, as of 1816, had been flourishing in at least six villages along the Upper White River near today's cities of Muncie and Anderson. They had received that tract of about 100 square miles from the Piankashaw branch of the Miami on condition that they never alienate it without permission from the Piankashaw. They grew corn, potatoes, turnips, cabbage, "pickles," beans, beets, carrots, watermelon, marsh-melon, and pumpkins. They also gathered and dried large quantities of wild blackberries to bake in their cornbread (considered a delicacy) and made large amounts of maple sugar.[38]

Back during the interval between 1794's Battle of Fallen Timbers and the War of 1812, the southern one-third of Indiana had been ceded to the United States government through six separate treaties.[39]

The final, northern one-third of the state was alienated *after* the War of 1812 through a succession of eight treaties, the last of them signed by Miami in 1840. The largest northern land alienation occurred through two 1832 treaties signed at Tippecanoe in the wake of the federal Indian Removal Act of 1830. Individual Indian leaders who signed land treaties were often granted land allotments which enabled they and their families to stay in place rather than move west of the Mississippi River with the rest of their people. White settlement being far advanced by this time, Indians were allowed to sell their final lands to merchants and real-estate brokers such as the Ewing brothers, who in turn sold them to speculators and farmers.[40]

But so much for Euro-Americans' desire number 3—the control of natural resources, including land. Each reader can drawn his or her own conclusions about whether Indians would have been better off had they practiced Euro-Americans' type of individual land ownership.

The fourth category of Euro-American desires deals with military and paramilitary cooperation. Between the 1794 Battle of Fallen Timbers and the War of 1812, Indians lost their military control over Indiana. Yet the year 1814 still saw United States officials seeking active Indian participation in military campaigns against the British. At the July 1814 treaty-making council at Greenville, Ohio, the Miami reaffirmed their neutrality in the War of 1812, but one of the U.S. treaty commissioners told them, "You have now come forward to take us by the hand; we are equally anxious and willing to take you by the hand, but you must take up the tomahawk and with us strike our enemies. Then your great father, the President, will forgive the past."[41]

Euro-Americans often justified the expulsion of Indians by characterizing them as less productive farmers.[42] Even would-be benefactors of Indians sincerely believed this, including the Moravian missionaries among the Delaware on White River near today's city of Muncie, and Quaker missionaries who lived among the Miami near Fort Wayne. Apparently they considered it part of their mission to convince Indian males to acquire draft animals and personally work in the fields,[43] which from the perspective of some Indians would deprive women, children, and the elderly of their productive contribution to tribal well-being.[44] Yet as late as 1834, many white farmers in Montgomery County (the site of Crawfordsville) depended on buying corn from their soon-to-be-expelled Wea neighbors. "So frequently did this occur," says a local historian, "that the section about the Wea received from them [the settlers] the appellation, 'Egypt,' which it has since retained."[45]

Chapter Three

Why They Came

> The opportunity which settlers faced in the [Old] Northwest comprised both the peasant's opportunity for a home and the gambler's opportunity for a fortune.
>
> —William N. Parker,
> "From Northwest to Midwest," p. 17

Thomas Jefferson spent the late 1780s serving as United States minister to France. In the spring of 1787, while his Virginia friend James Madison was organizing a convention to meet in Philadelphia, Jefferson took to the road eastbound from Paris and passed through the countryside of Champagne. There he noted "no farm houses, all the people being gathered in villages."

This perturbed Jefferson. "Certain it is," he wrote, "that they are less happy and less virtuous in villages than they would be insulated with their families on the grounds they cultivate."[1] Jefferson thought of large cities as sources of corruption and thought of self-owned family farms as the basis for a viable democracy—that much is well-known. Apparently he also wanted farm families to live directly on their farmsteads rather than clustering cheek-by-jowl in villages.

When Jefferson went to France in the summer of 1784, he had already recommended the grid pattern which Congress approved that year as the basis for sectioning the Northwest Territory in preparation for its occupancy by

Americans. At the corners of the grid's square land parcels, four farmhouses could stand near each other as a miniature hamlet, but, beyond that, further clustering of farm families was not likely. Indeed, even four farmhouses sharing a boundary corner turned out to be exceptional[2] because the incoming Americans chose their farmhouse sites not so much to be near their neighbors as to be near springs of water, fertile soil, trees good for "mast" production (the acorns and other nuts that hogs ate), or near a stream on which boats could navigate.[3]

The consequences of thus dividing the Old Northwest into square parcels can be estimated if we mentally visualize the Midwest laid out differently—such as how the old French settlement of Vincennes was laid out. Vincennes had been founded by the year 1731[4] and its French commandant divided most of its vicinity into "longlots" such as were common in France. In French Canada, longlots lined the banks of the St. Lawrence River and many of its tributaries. They also stretched along the Mississippi River between St. Louis and Kaskaskia.

At Vincennes, on the *west* bank of the Wabash River (across the river from Vincennes village itself) the French settlers built their farmhouses right on their longlots, and spaced them an average of about 355 feet apart. That was the width, anyway, of the longlots, whose length ran more than a mile. Meanwhile those who owned the more numerous longlots on the *east* bank of the Wabash lived mainly in Vincennes village itself, not on their longlots. Their house-and-garden parcels in the village averaged about 150 feet square—in other words, about one-half acre each. The houses were encircled by verandas and surrounded by profuse gardens.[5]

At first, in the 1730s, Vincennes was simply a fur-trading and military post. Nearby stood a Piankashaw Indian village. Soon, however, Vincennes became agriculturally self-sufficient and an exporter of farm products. The year 1746 saw 600 barrels of Vincennes flour sent downstream to New Orleans.[6] In 1767, after the French and Indian War, the new British claimants of the Northwest took a census and found 427 people at Vincennes, including 232 free inhabitants of all ages and of both sexes, seventeen Indian slaves, ten African–American slaves, and 168 "strangers." Most of those "strangers" (meaning non-inhabitants) were presumably fur traders and trappers.[7]

The Midwestern French excelled at gardening. Surrounding their homes they grew flowers, fruits, and vegetables in profusion. Most of them also farmed. As of the year 1767, the settlers at Vincennes produced an average of forty-nine bushels of grain per person, almost all of it wheat and corn. Three grain-grinding mills were located in the area. Not every year saw the settlers produce enough wheat to sell flour to merchants at New Orleans, but often they did achieve that. And they grew some of North America's finest tobacco.[8]

Vincennes and Illinois settlers were motivated to produce so much because a major market for food existed at New Orleans and elsewhere in Lower Louisiana. Twice each year a government-sponsored convoy of flatboats and pirogues was sent up the Mississippi loaded with imported merchandise. Three or four months were required to pole and row the convoy to Kaskaskia and its sister settlements. There the merchandise was unloaded and warehoused. Then the boats were loaded with farm produce for the journey back downstream to New Orleans, which took only three to four weeks.[9]

Vincennes settlers (again, this was in 1767) owned an average of 5.8 head of livestock animals per person, excluding chickens and other fowls. They did not grow much hay since their livestock subsisted all year round by foraging, doing so during the summer months on the village commons, which held about 5,000 acres (exceeding seven square miles) and then foraging during the winter months on the open unfenced fields, where crops had grown during the summer. This, at any rate, was the custom on the more-populated *east* bank of the Wabash where Vincennes village stood. Precise sections of the fencing that surrounded the Commons, and surrounded the overall "prairie" of longlots, were assigned to each household for upkeep, and wood for those fencing purposes was brought downstream on the Wabash.[10]

On the west bank, by contrast, where the French settlers lived directly on their longlots, they apparently fenced in their individual holdings and did *not* use those arable lands as a winter commons for livestock. But most Vincennes settlers lived on the river's east bank and practiced traditional French open-field farming and common pasturage.

The scholar Carl J. Ekberg notes that, as the French grew accustomed to farming Midwestern bottomlands, they "began producing relatively more maize and relatively less wheat. They fed [the maize] to their animals and to their slaves, they made whiskey from it, and they even debased themselves by eating corn bread."[11]

Two oxen pulled the moldboard plows (mounted on ten-to-twelve-foot carriages) that the French used to plow their longlots each spring. Most farm families personally owned an ox team and a plow. After plowing and harrowing, spring wheat was sowed on some of the land and harrowed slightly into the soil. Corn was planted in rows rather than Indian-style in hillocks. The oxen pulling the plows and harrows were reportedly not yoked across their shoulders, as we might visualize. Instead, a stick was tied in front of the base of their horns and that stick was what the pulling straps were hitched to.[12]

Meanwhile, one hundred and seventy miles up the Wabash River, the French post at Ouiatenon (near today's city of Lafayette) remained so focused on the fur trade that food shipments were sometimes sent upstream from Vincennes to supply Ouiatenon's residents. Ouiatenon had been founded in 1719 or 1720, about ten years before Vincennes, but its economy failed to diversify until after the French period ended. Vincennes as early as 1733 was exporting 30,000 deerskins annually, but both Ouiatenon and Miamitown (near where Fort Wayne now sits) exported more than that.[13] Vincennes products surely traveled with the government-sponsored convoys that plied between the Illinois settlements and New Orleans (see page 31 above) but plenty of private flotillas likewise plied up and down the Mississippi.[14]

If France had won the French and Indian War, or at least had fought the British to a draw and had maintained a claim to land east of the Mississippi River, fewer Anglo-Americans would have arrived from the east and many of those who did come might plausibly have adopted the French longlot farming system. Had that happened, today's Midwest might look different than it actually does. Ekberg thinks that the isolation which American farm families experienced on their square farm plots helped to foster the exceptional violence of midwestern Americans in the early 1800s.[15]

Granted that by the year 1783, when the Treaty of Paris acknowledged American independence, it might have been hard to resist the tide of Anglo-American land tenure patterns even if that year's treaty had reassigned the Midwest to French rule. After all, France and the new United States were allies by then, and American settlers, along with their customary dispersed farmsteads, would presumably have been accepted into a French Midwest.

But even so, fewer Americans would have come west had the Midwest stayed under French rule. For surely the French would have offered no special inducements to Anglo-American settlers—something which both the British and then the American government did offer them.

In other words, the westward push from British America's original East Coast settlements was not simply a triumph of laissez-faire free enterprise. Actually the colonial governments of the 1700s offered many inducements to potential settlers: quitrents were remitted, land was often granted free, and some settlers were even paid cash bounties for moving west. As far back as the 1720s the Board of Trade in London had begun fostering such inducements so that white Protestants of whatever stripe, even Scots-Irish from Ulster and German-speakers from the Continent, would settle in British America's backcountry all the way from Nova Scotia to South Carolina.[16]

Many Anglo-Americans looked upon Protestants of such stripes with repulsion. Benjamin Franklin's antipathy to German settlers is well-known. The genteel William Byrd of Virginia disliked the Scots-Irish even more—"who flock over thither in such numbers," he complained, "that there is not elbow-room for them. They swarm like the Goths and Vandals of old, and will over-spread our continent soon."[17] But regardless of elitist attitudes among some Americans, the British government made major efforts before 1750 to have white Protestants settle "deep in the contested interior of North America," as one historian puts it—which indicated, he adds, "an unprecedented aggressiveness in British plans for territorial conquest."[18]

Colonial Virginia had already moved in that direction in 1701 when Queen Anne's War began and Virginia's House of Burgesses passed an act "for the better strengthening of the frontiers." Realizing that it would take thicker frontier settlement to fend off the French and their Indian allies, colonial Virginia decided in 1701 to give actual settlers small tracts of land (usually several hundred acres) virtually free of charge. This was a sharp contrast to the high land prices that the Penn family was enforcing in Pennsylvania and it drew Scots-Irish and German land-seekers southward into the backcountry behind the Blue Ridge.[19]

And yet, after they defeated the French and asserted rule over trans-Appalachia, the British suddenly reversed that policy and tried to *stop* settlers from heading west. That in turn helped provoke the American Revolution, which then turned western policy topsy-turvy again. As another historian says, "In a crucial moment, when the direction of the Revolutionary war hung in the balance, the Continental Congress offered sweeping power and authority to Western settlers who had previously been regarded as outlaws in the British imperial scheme. . . . For strategic as well as ideological reasons, the states and the Continental Congress empowered westerners and thereby validated their pursuit of free land. . . . To retain westerners in its interest, the United States government had to define its territorial ambitions broadly and pursue them aggressively."[20]

In other words, the embattled U.S. revolutionary government decided to adopt a laissez-faire policy toward the West of that day. Meanwhile, some westerners were already practicing such a policy toward each other. The surveyor of George Washington's western lands had written to Washington in 1772, "As soon as a man's back is turned, another is on his land. The man that is strong and is able to make others afraid of him seems to have the best chance as times go now."[21]

Today, with our benefit of hindsight, we can see that a less individualistic pattern of settlement might have helped sustain social bonds and kept Americans at peace with each other. In the French midwestern settlements prior to 1763, for instance, violent crime and even violent altercations had been virtually unknown among the French settlers themselves. Only one such village case ever reached the local legal system—a case of name-calling. But that peaceful milieu changed dramatically as American settlers began arriving.[22]

At least part of this change derived from differences between French communalism and American individualism. The economic historian William N. Parker says that "Midwestern individualism, as it showed itself in farming and in business achievement, was rooted in the family organization brought into the region and reinforced by the conditions of rural settlement on isolated farmsteads."[23] The "family organization" that Parker means is the nuclear family. A French merchant who lived in Louisiana at the time said Anglo-Americans "will bring with them, in[to this] free and peaceful country, the discord and disunion of families through lawsuits and taxation. Lawyers, sheriffs, and constables will come crowding in here."[24] During the French era, by contrast, the legal system had emphasized reconciliation, which was underwritten by the paternalistic authority of both judges and priests.

Now in our own day of the early 2000s, after 230 years of laissez-faire land policy, the Midwest may be inching back toward old-time French-style community rights. Some lawyers are now suggesting that health-conscious limits be placed on the extent of private land-use rights. The influential University of Illinois law professor Eric T. Freyfogel says local communities "have not only the power but also the duty to reshape landownership laws so that they reflect evolving community sentiment, banning destructive resource-use practices, whether to water, land, or other elements of nature's fabric. They need to set appropriate standards for landowner behavior and to take action when those standards aren't met."[25]

The community-first practices of the Midwest's early French were not so different from those of the region's Native Americans. Both shared the idea that land uses should be regulated by membership groups which had the power to sanction or dispossess individuals whose practices broke social norms.[26] But for 230 subsequent years, down to today, *private* control over land use has been virtually absolute in the Midwest—to the point that hyperindustrialization (including the industrialization of farming) is now driving local governments to reinvent limits on how far they allow private property within their jurisdiction to be destroyed or contaminated by property owners themselves. Recently in Pennsylvania, many local townships passed laws outlawing factory-scale hog farms and also laws that prohibit spreading sewage sludge as fertilizer on farmland. A 2004 magazine reported that "no new factory farms have been sited, and no new sludge has been applied in communities that passed these laws."[27]

Holding property owners thus accountable runs counter to many Hoosiers' personal dispositions. An attitude of laissez-faire individualism characterizes rural Indiana. As we have seen, that attitude was reinforced by the western policy of the new United States in the late 1700s, but it had already begun forming during the 1600s in the continent's eastern forests, where forebears of Indiana's Euro-American pioneers were creating farms by using the same methods as would later be used in Indiana. By the year 1850, a million farms would be hewn out of the vast forest that stretched from the Appalachians to the Mississippi River, and about 100,000 of those farms would be in Indiana.[28] An immense expenditure of toil, guided by acute application of know-how, would lead by 1850 to an era of family farming so bountifully productive that some have called the 1850s Indiana's "golden age."[29]

It *was* a golden age, and it culminated centuries of exertions—exertions which had actually started long before Europeans colonized America. Some of the strands in Indiana's Euro-American farm heritage went back at least to the fall of the Roman Empire. Euro-Americans' land *hunger* certainly went back that far.[30] A brief glance backward can put the trans-Atlantic migration of millions of land-hungry European families into perspective.

Back in the 300s A.D., when ancient Rome's economy was faltering and Germanic tribes were breaching its frontiers and starting to slice that old empire into bite sizes, Roman officials tried to guarantee the empire's tax revenues by outlawing mobility on the part of its free tenant farmers, the farmers who rented the land they farmed. That was enacted in A.D. 332. Forty years later, in A.D. 371, Rome reversed that tax policy and gave those tenant farmers total exemption from taxes, but it still required them to stay in place.[31] Meanwhile, with tribal Germans roaming the empire at will, long-distance trade kept growing less safe, and money thereby was growing less useful.[32]

One adjustment that people made can be seen in their economic contracts, which increasingly specified barter "in kind," two-way deliveries of goods and services. Combined with the worsening security conditions, this increased the advantages of controlling land. Security threats prompted many rural people to promise free deliveries of goods and services to the strongholds of local warlords in exchange for the warlords providing the whole neighborhood with armed protection. Later this was called the manorial system, and it became a defining trait of feudalism.

Economically, it all added up to retrenchment, but given the unsafe conditions in those Dark Ages, it was progress. Technological advances no longer garnished Rome's famous grandeur with more columned public buildings or spacious forums, or even any more road-paving projects. Instead, Europe's Dark Age technology advanced agricultural productivity—which incidentally had stayed stagnant throughout the entire Roman era from the 400s B.C. to the 400s A.D. (Ancient Rome had become capitalistic but it had not industrialized.)[33]

During northern Europe's Dark Ages, farm innovators started the shoeing of farm horses and invented the horse collar, the breast-strap horse harness, the deep-cutting share plow, the coulter knife that cut sod in that plow's path, and the moldboard that turned over the soil thus plowed. Dark Age farmers also put wheels under all that equipment to convey it through the fields. For harvesting grain, the scythe was invented to replace the backbreaking sickle.[34]

After several centuries, security conditions brightened a bit. The fearsome Magyars (Hungarians) finally settled down, as did many of the fearsome Vikings in spots south of Scandinavia, where they too mellowed. Overall European security began rebounding by about the mid-900s.[35] And as security improved, the manorial system of in-kind exchanges of goods and services was supplemented by renewed use of money. A few centuries later, Europeans discovered the Western Hemisphere and the investment of money in American ventures helped well-to-do Europeans gain large profits by exploiting New World raw materials and people.[36] (In Chapter Six, we will see such investments still occurring in the 1800s in the place by then called Indiana.)

Meanwhile, Native American wonder crops like maize and potatoes were introduced into Europe in the 1500s and led to a swell in Europe's population. In the 1600s, little Europe grew quite crowded, and that was several generations before industrialization started to take up the slack. Europe's population boom in the 1600s, fed partly by the planting of American food staples in Europe, made some Europeans better off while it impoverished others. Industrialization grew widespread only when large numbers of Europeans became so desperately poor they were willing to go down in coal mines or to toil fourteen-hour workdays in what William Blake called "dark satanic mills."[37]

Often, Europe's desperate paupers possessed an alternative. Rather than selling their labor for a pittance, they could take the risk of migrating to America on credit. Once in America, and once they had completed their years of indentured servitude, such former paupers sometimes grew "uppity" and tolerated few social controls, especially if they had sufficient money to "go their own way."[38] Going their own way, in fact, might eventually take them or their descendants to Indiana.

Chapter Four

Where They Stopped and What They Started

Eastern historians always refer to them as "poor white trash." The world had not seen their like nor will it again.
—Logan Esarey in *The Indiana Home*, p. 15

Land does not raise much in this place owning to the great immigration down the River. It seems as if people were mad to git afloat on the Ohio. Many leaves pretty good liveings here sets of for they know not where but too often find their mistake. . . . The Mississippi trade is open at this time and all the wheat, whiskey, bacon, etc., buying up by those concerned in it.
—Israel Shreve, letter from Forks of the Yough in western Pennsylvania, 26 Dec. 1789, quoted in R. Eugene Harper, *The Transformation of Western Pennsylvania*, p. 130

When Israel Shreve wrote the above in 1789, almost all of those would-be settlers who were setting themselves afloat down the Ohio intended to disembark *south* of that river, not north of it. That changed, however, after the 1794 Battle of

Fallen Timbers. Then the river's northern shoreline started to fill with settlers as well. In 1800, almost all of the Northwest Territory west of about 85 degrees longitude was separated from what soon would become (in 1803) the state of Ohio, and what lay west of that line was given the new name Indiana Territory.[1]

In 1804 the federal government opened a land office at Vincennes, the territorial capital located on the lower Wabash River. Three years later it also began making land sales at Jeffersonville, on the Ohio River in southeastern Indiana Territory.[2] After the War of 1812 cut short the life of Tecumseh and intimidated most of the Midwest's surviving Indians, Euro-Americans began pouring down the Ohio River from Pittsburgh with a mind to become landowners. One decision they had to make was how far down the Ohio to go. If they stopped in Ohio to seek land, they could by 1817 face prices of up to $50 an acre for improved land and as high as $30 an acre for unimproved.[3] If, however, they continued on downriver to Indiana they could find good land still available from the U.S. government for $2 an acre—land that was unimproved, of course.

By 1816 when Indiana became a state, hundreds of settlers had bought land at the government's Cincinnati, and then Jeffersonville, land offices in order to carve out choice farms along the Whitewater River. The Whitewater flows south for sixty miles just west of Indiana's border with Ohio.

The newcomers there soon learned, as others around the new state would also discover, that Indiana's best soil was far too rich for the main money crop of that early era, which was wheat. As a local historian reported, "Many of the [first] settlers sought land along the streams of [Franklin] county and ignored the higher lands atop the flats. [But] the bottom lands of the Whitewater valley were so rich in vegetable matter that wheat was not profitably grown for many years. The fertility of the soil, by successive cropping to corn, became so exhausted that wheat has now become one of the major crops."[4]

Likewise by that statehood year of 1816, other hundreds of settlers had already bought south*eastern* Indiana land at the Vincennes land office and were creating farms along the lower Wabash River.

Meanwhile, the interior hill country of south-central Indiana held almost no Euro-Americans as yet, although it too had been ceded by Indians and had been surveyed by the government. Much of its best land was still in federal hands and thus still bargain-priced at $2 an acre.[5] One family which did settle there was that of Thomas and Nancy Lincoln, Abraham Lincoln's parents, who arrived in the late fall of 1816 when Abraham was seven years old. Thomas claimed two eighty-acre quarter-sections of land near the headwaters of Little Pigeon Creek in Perry County, fifteen miles from the Ohio River. It was land Thomas had selected during a preliminary trip earlier that fall, when also he probably erected a rough shelter on the land. It is likely the Lincolns erected a cabin immediately after their arrival.[6]

Forebears of Thomas Lincoln had arrived in Massachusetts in 1637 and later had lived in New Jersey before living successively in Pennsylvania, Virginia, and Kentucky. Thomas was part of the sixth generation of his line to live in America. As for Nancy Hanks, her mother's family had been in America for five generations, beginning in 1668 as Tidewater Virginians. Thomas Lincoln's main reason for leaving Kentucky in 1816 was evidently a succession of land losses that he suffered through legal challenges against his deeds.[7]

Soon, in 1820, the government would lower the minimum price to $1.25 an acre, and would lower the minimum permitted purchase size from 160 acres down to 80 acres. That was also, however, when the government began demanding cash payment in full at the time of purchase, allowing no more land-buying on credit by paying one-fourth down and the rest over the next five years. But the banking system immediately responded to that new government demand for full payment in cash. Banks and land speculators streamlined their mortgage operations,[8] and land-buying, although somewhat abated, continued at a fairly brisk pace in southern Indiana.

Many incoming families were simply moving north into Indiana from Kentucky, and they had things easier than families who came from afar by boat down the Ohio River.[9] The boat travelers, as total newcomers to the frontier, had to bring along with them many more goods, at least if the entire family would be living on its new land from the "git-go." By contrast, many incoming Kentuckians had already spied out Indiana land on hunting trips and had prechosen a likely spot on which to settle. Then the father of the family might come in the springtime (perhaps with his brothers or his older sons) and cut the brush off an acre or two, burn it, plant corn on it, and let the corn grow while they lived in a lean-to or a three-sided pole-and-brush shelter and constructed a cabin. After harvesting the corn and putting most of it in storage, he would return to Kentucky for the winter, then bring his whole family to the new land the next spring.[10] Along with the family would come its dogs and also perhaps a few sheep to provide wool for clothing, although buckskin could serve in a pinch.

Today, all this might sound intimidating, but bear in mind that pioneering was not new to most of the people who did it. Most of them had been born and bred to it and knew what to do. Many of those pioneers, or their parents or grandparents, had earlier moved out of Pennsylvania or Virginia to North Carolina or Tennessee and then on to Kentucky. (Later, many of them, or their children or grandchildren, would relocate from southern Indiana to central or northern Indiana.)[11]

Emigrant Boat in which the Pioneers Went from Pittsburg to Kentucky

The men knew, and the boys were learning, how to appraise soil fertility and other land potential, how to fell trees in a chosen direction, how to build with logs and use wood in countless other ways, how to nurture and use livestock, clear land and plant it, cultivate, harvest, and preserve crops, how to fish, hunt, trap, tan skins, and preserve furs. They knew the lore of hound dogs, honey bees, fruit trees, maple sugar making, and dozens of other useful things.[12]

As for the women, they knew (and the girls were learning) how to keep house, grow a garden, milk cows, churn butter, raise sheep, geese, and chickens, cook and bake, preserve all kinds of food, spin thread, weave cloth and how to bleach, dye, and full it and make clothing from it, how to leach ashes to make lye, how to make soap and candles, buckskin, and dozens of other useful accessories. They also knew how to find, preserve, and prescribe herbs, how to transplant flowers, and how to manage a busy household.[13]

Their supply of tools was skimpy. The men's basic tools were their gun, knife, axe, and shovel plow. The women's were their pot and three-legged skillet (which with a lid and without its legs could perhaps double as a Dutch oven), their distaff and spindle for spinning, their loom for weaving, and their kitchen knife. A hoe was needed too but both sexes needed it equally. Logan Esarey says:

> they brought very little property—on an average a horse or "critter" for each family. Apparently about half, before 1816, came in wagons and of these latter at least one-half in ox wagons. Of those who came on foot or horseback, the average household property was a quilt or coverlet, a change

> of clothing, a pot and "spider" or three-legged skillet, an axe, hatchet, two or three steel knives, a hoe and a few other trinkets or trifles. To this outfit the wagoner added a few more tools, a plow, some pewter dishes, a box of valuables—the family "chist"—and some bed clothes. There were no traveller's guides and hence no regularity in the outfit. Each grown man, without fail, carried a rifle and with each family came one or more dogs. Milk cows, pigs, chickens, farm seeds and the like were brought by those who could afford them. Clocks, carpenters' kits and other tools were to be found here and there—one or two in a neighborhood.[14]

Logan Esarey wrote this about his forebears and their friends to emphasize their poverty. He says that "scarcely one in a hundred had half enough money to pay for his home."[15]

The work of pioneer farm-making has been described not only by Logan Esarey in *The Indiana Home* but also by Robert P. Duncan in "Old Settlers" and by Oliver Johnson in *A Home in the Woods.*[16] Both Duncan and Johnson grew up in a woods that is now Indianapolis, and Esarey grew up in Perry County in far south-central Indiana. Every family's expedients varied. Many spent their first winter *not* as yet in a cabin but still in a lean-to or three-sided shelter. If so, the fire for cooking and heating was placed beyond the open end of the shelter, which typically faced south.[17] Most families built a one-room cabin even if they did not yet own the land it was on, but other building efforts were often postponed until the public land auction—just in case some "low-down speculator" took a mind to outbid the actual settler.[18] Another common strategy among newcomers was to wait two or three years before buying land and meanwhile to live as tenant farmers in the vicinity of their planned homestead, renting the use of land from which they could expect successful crops. As tenants who were thus profiting from a few years of crops, such newcomers could save up more money for land-buying and meanwhile could soften the rigors of their first few western years.[19]

German-American pioneers in Indiana would try the hardest to clear their crop fields "smack-smooth," removing even stumps and large rocks, but other pioneers generally just deadened the trees by girdling them with an axe, which would kill the leaves on some common species in a matter of mere days, including on hickory, oak, sugar maple, beech, and yellow poplar. A stand of girdled trees was known as a "deadening." The rapidly withering leaves allowed sunlight through to reach the crops.[20]

After the initial settlement year, men and boys customarily cleared more land by deadening trees in July or August and then letting them stand two or three years before cutting them down. Beech trees, however, and sometimes other species could eventually be burned standing, especially if they had been deadened while their sap was up, since that made them more combustible later. If cut down, and if not needed for making fence rails or for building, the trees prompted a log-rolling and the month for this was May. Neighbors were invited, and the logs, which during the winter had been cut or cross-burned into lengths of twelve to twenty feet, were rolled close together and piled up for burning, each pile containing up to ten logs. The day's host was expected to provide a jug or two of well-aged whiskey to pass among the assembled crew, just as would be provided at a log-raising or any other neighborly project. Often the logs were immense, but enough men wielding handspikes could somehow carry or roll them together and pile them. Competition served to enliven all this effort. Sometimes back trouble resulted. Meanwhile, the women and older girls would be quilting at the cabin and preparing a feast for supper.[21]

As for the brush, it was best to cut that in the fall and burn it the next spring just before planting so the fertile ashes would not leach away but would fertilize the crops. Fertilization was also the motive behind burning the surplus logs in May. But if winter wheat was the crop intended to be sown, then the burning of logs and brush might be postponed until late August, just before the winter wheat was planted.[22]

It was all unsightly but it worked. If you picture the "cropland" that such methods produced, you can understand why a deep-cutting plow would not work there. Those new "fields" were still crammed with rocks, roots, and massive stumps. They were worked with a "jumping shovel plow," which jumped over obstacles thanks to having a cutter bar positioned right in front of the plow's iron shovel-point. In its narrow designs it was called a bull-tongue plow. One of its attractions was its cheapness; as late as 1833 a new shovel plowshare (that is, the metal part) was selling for just eighty-seven cents in Tippecanoe County.[23] A single horse or preferably a pair or two of oxen pulled that plow—preferably oxen because they walked only half as fast as a horse and thus minimized accidents. Shovel plows hardly plowed at all, but that was enough in forest soil. Mixing new ashes into the forest's acidic humus served to lower the acidity enough to raise corn, perhaps along with some small grains, hay, and vegetables.

At first in many areas, ginseng grew so thickly that by trading it to storekeepers the earliest settlers could finance their store purchases. "Ginseng grows in the bottoms to a perfection and size, I never before witnessed," said Samuel R. Brown in his *Western Gazetteer*, "and so thick, where the hogs have not thinned it, that one could dig a bushel in a very short time." That bonanza was soon depleted, however, through overharvesting by humans and the appetite of the razorback hogs.[24]

After ginseng was depleted, the foraging hogs themselves became the next "cash crop" of a typical pioneer farm. They were lean, long-legged critters with short pointy ears. They ran semi-wild and were called many names but answered to none—razorbacks, land-sharks, alligators, elm-peelers, wood-hogs, tonawandas, land-pikes, wind-splitters, hazel-splitters.[25] One historian emphasizes that they "quickly adapted themselves to their wild surroundings. Wolves sometimes killed young pigs but the mature hogs, especially the males, were formidable antagonists. They sometimes became so vicious and dangerous that in squads, they would even attack and kill a bear." Those hogs were "sorry-looking specimens" that became fully mature at age two or two-and-a-half simply by foraging on "forest products, which, when abundant, put them in condition for market. Sometimes, after being finished on corn a few weeks before Christmas, they weighed 200 to 250 pounds."[26]

Each fall the surplus hogs that were not needed for home consumption or for breeding purposes could be walked to a market town like Cincinnati, Ohio or Madison, Indiana (both located on the Ohio River) and sold there to a pork packer. In November 1834 a Brookville, Indiana newspaper editor estimated that 30,000 hogs had been driven through his town in three weeks' time, all heading toward Cincinnati.[27] The animals would walk about ten miles a day. Such drives might contain as many as 2,000 hogs each and they required about one drover for each 100 hogs. Occasionally much longer drives were also undertaken. In the 1820s and probably into the 1830s, razorback hogs sometimes were driven from Indiana all they way to the East Coast in droves of about 600 at a time.[28]

The cash crop *most* potentially profitable was wheat, but much of Indiana's soil was initially too rich to grow good wheat. Often it took from four to fifteen corn crops (depending on the soil) to lower the soil fertility enough to make the wheat stalks "head" properly and not "bolt."[29]

In most of Indiana, settlers planted wheat in the early fall (unlike the Vincennes French) and saw it start ripening early the next summer about June 20. Wheat thus ripened conveniently just when the corn crop could be "laid by," which meant that it needed no more cultivation (hoeing) because it had gained a sufficient advantage against the weeds to produce a good corn harvest in the fall. Thus, corn required hoeing for only six or seven weeks in May and June—and then in midsummer it was time to harvest the wheat. That was traditionally tackled with cradle scythes and was often a neighborly endeavor, again with competition enlivening the effort. And the same was true of attacking the *hay* crop—scything, binding, and stacking it—which usually started about August 1.[30]

August 1 was also the date when an observer in 1824 saw a ripe fifty-acre field of cotton "in full boll" in the Point Commerce part of west-central Indiana's Greene County. "It was a beautiful sight to behold . . . for the wide expanse of terrain was blanketed with solid white."[31] That field was a composite of individual plots of local residents. Cotton was raised quite extensively in the early 1800s in southern Indiana prior to factory-made cotton cloth appearing in the stores.

And also by August 1, some of the pioneers' corn crop was probably reaching what the Indians called "green corn" and the settlers called "roasting ears" or "in the milk." But the main corn harvest came later, in autumn, and the signal for that was the first frost. Many of the cornstalks were cut by hand with corn knives (long machetes) and were piled upright in the field in shocks with the ears of corn still on the stalks. After curing that way, the ears were usually removed before livestock were allowed into the cornfield to feed—although that was arguably a waste of time if the corn was meant anyway for hog feed.[32] The rest of the corn crop (the part that was kept off-limits to livestock) was left standing where it grew and eventually its ears were removed and hauled to the barn or corn crib to dry further. Shucking the corn was postponed until midwinter, when a neighborly "corn-husking bee" could enliven a long evening.[33] By Hoosier custom, boys who found an ear of red corn could kiss the girls who were present.[34]

Back in Chapter One we saw that midwestern Native Americans' diets consisted about 24 percent of corn around A.D. 1000, and that then by A.D. 1300 the corn content in their diets rose to about 70 percent.[35] I don't know any estimates of corn's percentage in the diets of Euro-American pioneers but "at least half" might be a safe guess. Many pioneers ate corn morning, noon, and night, usually with pork in some form—although, early on, venison and wild turkey also furnished protein.[36] Pork came in many forms, like today. The pork (and the lard too) from old-fashioned razorback hogs tended to be oily and soft—especially if they ate mostly acorns and other nuts as they usually did. Even if the razorback hogs were allowed to consume corn for a month or two before they were slaughtered, their meat and lard often remained oily and soft.[37]

As for human consumption of corn, that was easiest to do by eating dodger, pone, or johnnycake. All three called for cornmeal and it did not need to be ground very fine. Dodger is simply cornmeal mixed with a little water and salt, then packed into a ball and baked in a Dutch oven that is buried in hot ashes at the bottom of a fireplace. Corn pone is similar but with some milk and yeast added so it rises and resembles a loaf of bread. As for johnnycake (also called hoecake) that is just cornmeal mixed with shortening—such as lard, bear grease, or butter—and then baked on a board of wood. A common diet was johnnycake and pork for breakfast, pone and pork for dinner (nowadays called lunch), and corn mush and pork for supper. The mush, of course, was just cornmeal boiled in a pot.

The pioneers could also eat corn without having to grind it. First, wood ashes were leached to make lye and then the corn was soaked in diluted lye, making hominy, which could either be boiled or fried depending whether you wanted hominy mush or hominy cakes.[38] Making a batch of hominy took about a gallon of wood ashes and three quarts of white corn kernels. First the ashes were boiled briefly in about two gallons of water. Then the lye water was leeched out of that and the ashes were tossed away (preferably on the garden). The lye water was next used to boil the three quarts of corn for about twenty-five minutes, until the corn kernels' outer layer started to peel off. The corn was then scooped out of the lye water and put in a pan of clean water, where it was rubbed to fully remove the kernels' outer layer. After rinsing the softened kernels a few more times, they were simmered about an hour until they were double their original size and quite soft. They were then ready to eat—either in a bowl as hominy mush or else fried as hominy grits.[39]

In the fall, surplus supplies of pork and corn, and also of wheat and the other small grains, along with other products, were gathered together, some to be shipped downriver soon if the fall rains put enough water in the rivers, but most to await shipment downriver on the high water of springtime. The next spring, in March or April, those goods were loaded on flatboats to be floated down the swollen rivers to New Orleans. Riding the high waters of springtime, the trip downriver to New Orleans could take as few as four weeks. For flatboats starting from *central* Indiana, six weeks was considered good time to New Orleans.

As of 1820, only 12 percent of the crops grown in the Old Northwest were yet exported from the region. Almost all of those went downriver to New Orleans on flatboats.[40] In the early 1820s the Ohio Valley's total downriver exports required only about 3,000 flatboats to carry them all, but by the late 1820s Indiana alone was sending downstream an estimated one thousand boatloads of produce per year. About 300 of those flatboats were loaded directly on the Ohio River and another 300 or so came down the Wabash River, including many that started on the White River. Down the White River's East Fork, Lawrence County alone sent almost forty boatloads yearly, and down the White River's West Fork, Morgan County sent almost as many.

After unloading in New Orleans, or perhaps a bit further upstream, most of the boatmen returned home on foot—anyway until the 1837 financial crash reduced the price of steamboat tickets. Sometime in June the boat crews would arrive back in Indiana. The flatboat captains, however, often came back north on steamboats, bringing trade goods for storekeepers (who in fact might include themselves).[41]

Based on the total contents of the 152 flatboats that passed Vincennes in the spring of 1826 en route toward New Orleans, Indiana's annual river exports in the late 1820s already exceeded 1.6 million bushels of corn, 650,000 barrels of pork, 65,000 hams, 16,000 live cattle, 65,000 pounds of beeswax, 23,000 venison hams—and assorted hogs, chickens, oats, cornmeal, etc.[42] The 1826 Vincennes count apparently omitted barrels of whiskey, but they too were going downriver. A later record indicates that fourteen and a half million gallons of whiskey were sent downriver from Cincinnati alone in the year 1850.[43]

By taking these rough figures and calculating per individual person, we can get an impression of the "micro" significance of Indiana's downriver agricultural exports. Since the 1830 census reported that 344,508 people were then living in Indiana, the state as of 1826 probably held about 300,000 people. Thus the state's downriver exports in the late 1820s, if calculated per state resident, amounted roughly to five and one-third bushels of corn and just over two barrels of pork, along with other products in much lesser amounts.

Some records also survive of individual flatboat loads, such as this Indiana cargo that floated to New Orleans that same year of 1826: eighteen barrels of whiskey, seventy barrels of oats, 8,000 pounds of bulk pork, seven barrels of pork, 300 barrels of corn, and fifteen barrels of corn meal. In 1830 a flatboat from Leavenworth on the Ohio River set out with 2,000 pounds of bacon in bulk, 300 bushels of oats, twenty-nine hogsheads of tobacco, forty kegs of lard, thirty-five barrels of whiskey, thirty barrels of flour, and thirty barrels of corn meal. And in 1837 one Indiana flatboat took simply ninety tons of hay. The largest flatboat load ever recorded at New Orleans totaled 180 tons. It arrived in 1850 and consisted of 160 barrels of lard and 160 tons of lard in bulk. As of the 1846–47 season (from September 1846 through August 1847) the number of Indiana flatboats reaching New Orleans was 764, and some others had ended their voyages at Memphis, Vicksburg, and Natchez, which had become viable trade ports by the late 1820s thanks to the cotton boom in Mississippi, Alabama, and western Tennessee.[44]

The result of so many loaded flatboats arriving in New Orleans each spring was called the "Wabash Glut." It so drastically reduced New Orleans prices that some late-arriving boatloads were sold at a loss. At least one entrepreneur found in 1839 that he could not sell pork in New Orleans at any price and he dumped his whole shipment in the Mississippi River.[45]

A common way of financing such shipments was explained by the Madison, Indiana banker James F. D. Lanier, who made his first fortune as follows:

> We purchased largely bills drawn against shipment of products to [New Orleans]. As these bills were about to mature, it was my custom to go to New Orleans to invest their proceeds, and such other means as our bank could spare, in the purchase of bills drawn in New Orleans upon shipments of produce from thence to the Eastern States. The proceeds of the latter bills, at their maturity, supplied us amply with exchange for our western merchants, in payment of their purchases of merchandise. In this way we were enabled to turn over capital several times each year, and at a good profit.[46]

And of course not just products for export were grown. In fact, a family's own subsistence held priority.[47] In some of the pioneers' diets, potatoes were almost as important as corn. Logan Esarey chronicles the annual vegetable-planting sequence that women and children carried out. It began in March with the indoor planting of seeds to grow "starts" for cabbage, tobacco, and pepper plants (including hot red peppers). By mid-April came the outdoor planting of lettuce and radishes. Peas, bush beans, and garlic also went into the ground fairly early, but pole beans were most easily grown in the cornfield and thus were not planted until the new corn shoots had almost a six-inch head start, enough so the cornstalks would later support the bean vines. Both beans and potatoes would be planted at several times, some for early eating but the last planting to furnish winter provisions. Whenever possible, potato planting was started on Good Friday.

Cucumbers, squash, and pumpkins were planted after springtime's presumed last frost and their spreading vines helped to mulch the rest of the garden. The pepper starts were also transplanted out in the garden after the presumed last frost, including those hot red peppers. Turnip seeds were planted about June 1, and turnips were mostly eaten raw like apples in the fall and spring.[48]

Apples, in fact, were not really common in most of Indiana until later, notwithstanding legends about the exploits of Johnny Appleseed, who for many years operated a nursery near Fort Wayne.[49] Upland Southerners liked apple cider no less than Yankees did, but that craving could be met with just a few apple trees. And as late as 1820, a New England woman living at Lexington, Indiana, only a dozen miles from Madison on the Ohio River, lamented that apples were unavailable there.[50]

Around pioneers' cabins, gourds and flowers were often planted, some of the flowers being transplants from the woods. Small gourds were grown to use as drinking cups and ladles, whereas large gourds were grown to use as storage containers. Finally, the settlers waited until fall to plant large "potato onions." Those were covered with straw to protect them from the weather, and next spring they became the first vegetables available for eating.[51]

Indiana's early farm families did not just farm and garden, of course. They also hunted, fished, and gathered wild herbs, nuts, and fruits (including berries). Usually their food supplies were plenteous.

Variations in all these agricultural exertions were countless. In the state's far southeast, in what became Switzerland County, numerous French-speaking Swiss settlers began arriving in 1798. They specialized in raising grapes and making wine.[52] Logan Esarey relates that there, in the 1820s on the Ohio River "at Vevay, the Dufours and Schencks, known by their products throughout the east, loaded their annual fleet with wine, hay, straw hats, and other produce of this energetic French community."[53]

Even less typical, despite their English roots, were the Shakers. In 1817 the New York State Quaker Thomas Dean and a party of Native Americans visited the Shaker commune at Buseso Prairie, a bit north of Vincennes on the Lower Wabash. Thomas Dean and his Indian friends learned that those Shakers were starting more communes in Illinois, and also learned that the Shakers had fervently supported the aspirations of Tecumseh and his brother Tenskwatawa, the Prophet.[54]

The most populous exception to Hoosier individualism began in 1815 with the arrival from Pennsylvania of several hundred utopian Germans, adherents of the millenarian preacher George Rapp, who came in person to oversee everything. Like the French Swiss settlers and the Shakers, these communal Germans were highly entrepreneurial farmers. By 1819 they had achieved virtual self-sufficiency and were selling over $12,000 worth of their farm products annually. The Rappists' trade stretched far and wide from their village of New Harmony on the Lower Wabash. Their New Harmony store opened a branch at Vincennes and two branches in Illinois towns, and their aggressive commercialism affected everyone in the vicinity. At harvest time, the whole Rappist community would march out to the fields together led by the community band, and the pace of their grain harvest could reach 100 acres a day.[55]

Such Rappist festivities were more formalized than the festivities of upland southerners—the "butternuts" who composed most of Indiana's early settlers and whose log-rollings, cabin-raisings, and corn-huskings were not necessarily orderly but were famously festive.[56] Called "butternuts" because they extracted their typical clothing dye from butternut shells, those folks constituted about 90 percent of Indiana's Euro-American population before 1830. Historians who harbored a Yankee bias used to claim that progress was postponed until Yankee in-migrants started arriving in significant numbers in the 1830s. Meanwhile the butternuts supposedly clung to tradition and resisted progress. One historian claimed that the "Yankee invasion" which followed 1830 "was probably the most dynamic thing that played upon the social genesis of the Old Northwest or any other portion of America."[57]

Yet a statistical study of Old Northwest farm households has shown that (to quote it) the "southern-born farmer . . . was no poorer than the Yankee-born farmer."[58] And what above all made Yankee schemes viable was the emergence of the corn belt, which was started by southerners.

Chapter Five

Origins of the Corn Belt

One for the blackbird, two for the crow;
One for the cutworm and two to grow.
—Early corn-planting rhyme

Corn is the great staple of the State. It is easily cultivated and almost every farmer has from 20 to 100 acres. A single hand can prepare the ground, plant and attend to and gather from 20 to 25 acres, according to the state of the ground and character of the season.
—*The Indiana Gazetter* (1850), quoted in Thornbrough, *Indiana in the Civil War Era*, pp. 369–370

The corn belt emerged from a method of livestock fattening that by the 1760s had become systematized in some of the long, fertile valleys which lay nestled between Virginia's Blue Ridge Mountains.[1]

Native Americans in pre-contact Mexico had long since developed maize through selective breeding. Other Indian contributions to creating the corn belt included the use of fire for land-clearing and the cultivation of corn in bottomlands which later, after their appropriation by Euro-Americans, became "hearths" of the corn belt.[2] But the corn belt was not simply about growing corn. It was about growing corn to fatten both locally born and non-locally born livestock for commercial sale. The procedure that the corn belt's pioneers used was summarized by one of their sons:

> On the Scioto river much of the land was owned by immigrants from the south branch of the Potomac River, Virginia, where the feeding of cattle had been carried on for many years in a manner peculiar to that locality, and which materially differed from the mode practiced in Pennsylvania or further north. The cattle were not housed or sheltered, but simply fed twice a day in open lots of eight or ten or more acres each, with unhusked corn with the fodder, and followed by hogs to clean up the waste and offal.[3]

This was a system brought into Kentucky and the Midwest by cattle-fattening specialists whose families practiced it particularly in western Virginia's Hardy and Greenbrier counties.[4] The *largest* concentration of cattle-fattening during the colonial era occurred along the South Branch of the Potomac River in Hardy County, Virginia around the town of Moorefield. In the 1780s some of the cattle specialists there began acquiring purebred bulls that had been imported from England, and that was also when some of them (with names such as Patton, Renick, McNeill, and Van Meter) began bringing their cattle-feeding system and some of their best animals to certain favorable locations west of the Appalachians, notably to the Bluegrass region of Kentucky—also bringing along with them use of the corn shock, which they had devised to preserve corn in the field for use as winter forage.[5]

How their cattle-feeding system worked has been explained in greater detail by Robert Leslie Jones:

> After the stock cattle, usually four-year olds, were acquired, they were kept for the time being on whatever aftergrass or other good pasturage was available. Neither then nor at any time were they given shelter except natural windbreaks. During the feeding season—which lasted from about the first of November to the middle of February for the first animals [to be] driven [east to market], and to the middle of April for the last ones—the feeders, in herds of about a hundred, were turned into a succession of feedlots, a "fresh" one at each feeding, morning and late afternoon. Every day the hired men would haul as much corn as might be required on a big sled or a low feed wagon and distribute it, not in feed boxes (which were a rarity till after the Civil War), but in small heaps over the field. This was, in accordance with South Branch practice, unhusked corn—not mere blades and stalks—which had been standing in the stock since cutting time. When the feeders were put into their next feedlot, their places in the first would be taken by an equal number of stock cattle, who would eat as much as they could of what was left on the ground. Then, when the feeders were moved on to a third lot, and the stockers to the second, hogs would be turned into the first to

> consume what the cattle had mangled, trampled into the mud, or failed to digest. The limited number of feedlots meant that the first group of cattle would shortly be back to the first one (still called "fresh," however), following the hogs. . . . When a cattle feeder had plenty of first-class pasture, he could buy three-year-olds, corn feed them as stockers following feeders for one winter, graze them throughout the next growing season and into the fall, and then fatten them in the feedlots. This was how the best beef was produced.[6]

Practitioners of this system brought cattle west from Virginia despite their realization that western markets barely existed as yet. They intended to expand their herds in the West and then drive them back to their old homes in Virginia, where their weight could be re-bolstered before they were driven on further east to market cities like Philadelphia, Baltimore, and Alexandria. The first cattle drive east from Ohio's Scioto Valley was achieved in the spring of 1805 by George Renick with a herd of sixty-eight cattle. After he stopped to see relatives at Moorefield, Virginia and to sell twenty-one of the cattle there as stockers, he drove the rest of the herd to Baltimore and successfully sold them as fat cattle.[7]

At first, in the early 1800s, the Renicks and others who had brought their cattle system west had to "go looking" in order to acquire an adequate supply of grazed cattle ready for fattening. In 1810 George Renick's brother Felix, along with a friend from Chillicothe, traveled via the Natchez Trace to northern Mississippi and western Tennessee where they bought 500 lean cattle from Chickasaw Indians. By 1816, some cattle were being driven east from Missouri to Ohio, and in 1819 Felix Renick and his nephew William Renick traveled to Missouri to try to arrange a regular supply. It took another ten years, however, for the cattle supply from Missouri to Ohio to become regular.

William Renick, meanwhile, became active as a cattle buyer closer to home. By the mid-1820s, says Robert Leslie Jones, he "was 'picking up' cattle for his father [George Renick] over a wide area in Ohio and Kentucky. He ranged over the part of Ohio south of the fortieth parallel (which is slightly north of the National Road) from the western border of the grazing country along the Little Miami to the midst of the hill counties between the Scioto Valley and the Ohio River. In Kentucky he bought not only in the Bluegrass but sometimes also in the Green River section."[8]

Other Scioto Valley Ohioans were acquiring cattle from Illinois to fatten in their Ohio feedlots. In 1835 a young man named Benjamin Franklin Harris was sent west to south-central Illinois to pick up a herd for fattening in Ohio. He "later recalled how they had to swim the Kaskaskia River at Vandalia, Illinois, the Wabash at Attica, Indiana, then drove through Muncie, Springfield (Ohio) and Columbus, where they swam the Scioto." The young man also took fattened herds on east *from* Ohio, which "required swimming the Ohio at Wheeling before crossing the mountains to Cumberland, Hagerstown, Gettysburg, and Harrisburg, finally reaching the point of sale, Lancaster, Pennsylvania."[9]

Meanwhile, the Scioto Valley as a cattle-feeding center became one of the "hearths" from which its style of corn-and-livestock operation soon started to spread, eventually becoming the corn belt.[10] The geographer John Hudson is explicit that "the Corn Belt, from its inception until the late 1850s, was a cultural region, rooted in Upland Southern culture."[11] Prior to the railroads starting to mix cultures in the Midwest willy-nilly, says Hudson, the corn belt was an extension of the Upland South; but "by the late 1870s it would no longer be possible to speak of the Middle West in terms of population origins based on early settlement by one regional culture rather than another. Railroads effectively homogenized migration flows," and the Midwest became "a sort of regional melting pot."[12]

Already by about 1810, five main "hearths" of the future corn belt existed. Documentation links families from western Virginia's Hardy and Greenbrier counties to the three northernmost of those: the Bluegrass section of Kentucky, the Scioto Valley of south-central Ohio, and the Miami Valley of southwestern Ohio. By the time the first U.S. agricultural census was taken in 1840, specialized livestock fattening was also being extensively practiced in the Whitewater Valley of southeastern Indiana and in a few locations further west.

What sort of corn were they using? Well, it must have been gourdseed corn, the kind customarily used by settlers who came from the Upper South. Unlike the "flint" corn that was brought west from New England, New York, and Pennsylvania, gourdseed was soft enough for livestock to chew without the kernels first being cracked. Gourdseed also had the advantage of outproducing flint corn, but unfortunately gourdseed did not mature until fall and any exposure to a bad freeze might prevent germination of its kernels. Freezing did not hurt corn for use as food (including as cattle food) but freezing did nonetheless matter, since some of the kernels had to be planted the next spring as seeds. If the planted kernels did not germinate, farmers often filled the gaps in their fields by hastily replanting with seeds of the faster-ripening flint corn. Unintentionally, cross-pollination then created the modern "dent" corns by combining the early ripening of the Northeast's long, thin, straight-rowed flint ears with the chewable softness of the Upper South's short, stocky, crooked-rowed gourdseed ears. Eventually, dent overwhelmingly became the corn of the corn belt.[13]

In the 1840s, the fattening of livestock for market spread rapidly westward from Ohio. Using the U.S. agricultural census, John Hudson sets forth a two-part definition of a corn-belt county. It was a county that (1) had a relatively large livestock population, and (2) grew enough corn to fatten a major proportion of its livestock for sale to meatpackers or other merchants. If a county grew at least 7.5 bushels of corn per acre of its total amount of "improved" farmland, and if it additionally grew at least 18.5 bushels of corn per head of its resident cattle and hogs, Hudson counts it as part of the corn belt.[14]

Counties growing that much corn did not *necessarily* hold large numbers of livestock, but, using the census, Hudson shows that growing corn for cash sale (rather than for feeding livestock *in situ*) did not start in the Midwest on a large scale until the 1850s. And from then until at least 1920, says Hudson, such cash-corn farming remained concentrated mostly in the Grand Prairie region of east-central Illinois. Hudson's criterion for labeling a county a large producer of cash-sale corn is that its corn harvest exceeded sixty bushels for each cow or hog in that county.[15]

As of 1840 (which was when the U.S. agricultural census started) on the basis of Hudson's two-part criterion, the corn belt consisted of only five counties in Tennessee's Nashville Basin, five counties in Kentucky's Bluegrass section, two counties in Ohio's Scioto Valley, three in Ohio's Miami Valley, one county in Indiana's Lower Whitewater Valley (Dearborn County), one in Indiana's middle Wabash Valley (Montgomery County, the site of Crawfordsville), and two counties in northeastern Missouri.

But by ten years later in 1850, approximately 160 midwestern counties already qualified by Hudson's definition, including all of north-central Kentucky, all of southwestern Ohio, most of the central third of Indiana, all of the central third of Illinois (except one county), plus sixteen counties in eastern Iowa and twenty-three scattered around Missouri—a total of about 160 counties even with Tennessee excluded. In Indiana by that year (1850), thirty-eight counties were already part of the corn belt.[16]

By 1860, a few more Indiana counties were added to the corn belt, but its main growth by then was occurring in southern Illinois and across southern Iowa and northern and western Missouri.[17] A historian of Illinois and Iowa is explicit that there too, as in Ohio and Indiana, "feeding corn to cattle and then using hogs to salvage the waste was a successful commercial agricultural pattern."[18]

Thanks to corn, Indiana as a whole was remarkably self-sufficient—to the extent, anyway, that its inhabitants were willing to eat mainly corn and corn-fed animals. The U.S. census found that by 1859 Indiana farms were harvesting over 53 bushels of corn for every man, woman, and child in the state including urban dwellers.[19]

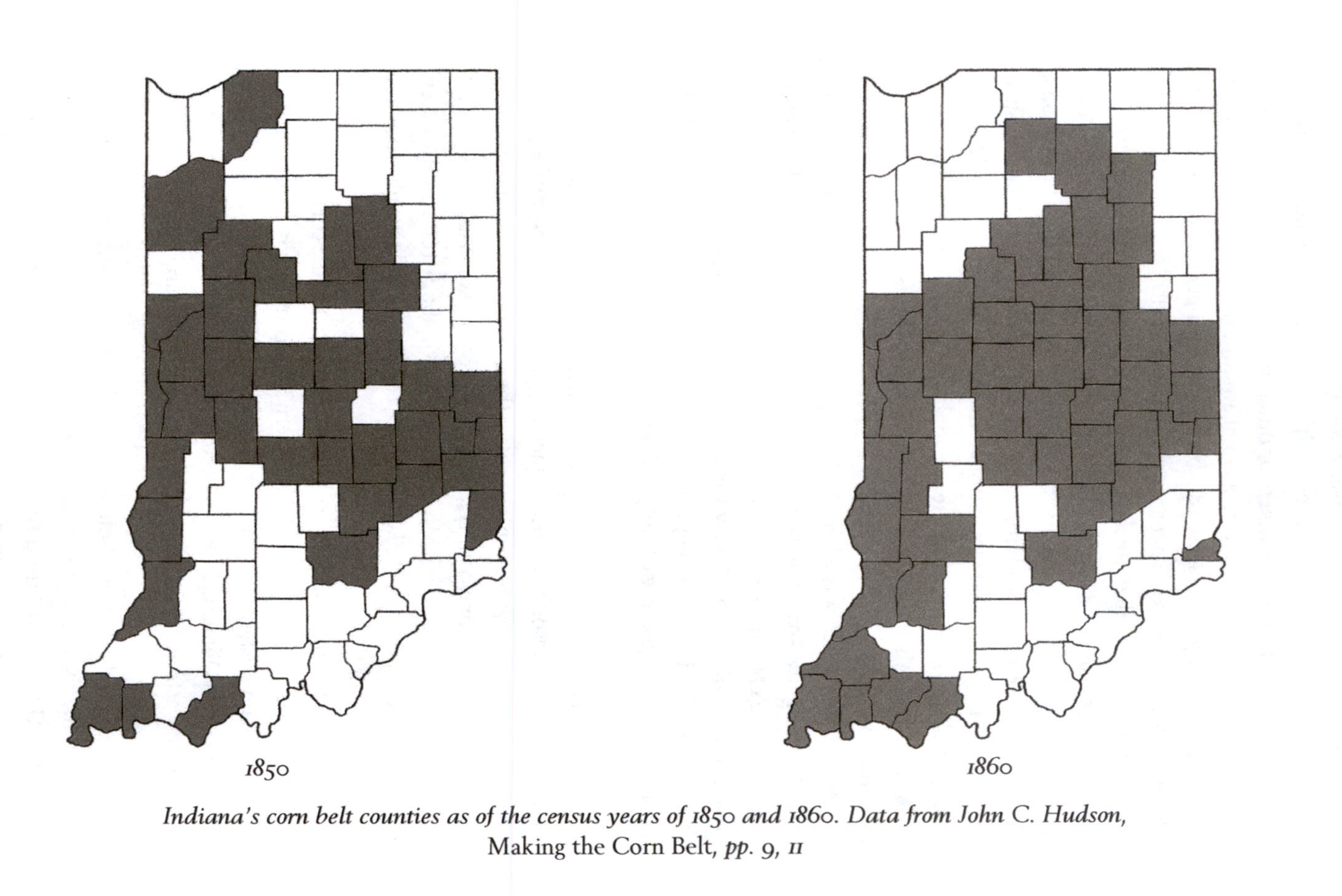

Indiana's corn belt counties as of the census years of 1850 and 1860. Data from John C. Hudson, Making the Corn Belt, *pp. 9, 11*

Where was it all growing? Throughout the state, of course, but the heaviest concentration grew in the Upper Wabash Valley, especially around Lafayette. As we have seen, the French who created the Ouiatenon outpost there (founded in 1719 or 1720) were content to trade furs; they did not always farm enough to even feed themselves. But as of 1718, five villages of Wea and Piankashaws had flourished in that vicinity with a total population exceeding 1,000. They had raised corn, melons, and pumpkins while buffalo grazed nearby.[20] Later that area's reputation for maize production reached the ears of Christopher Gist whilst he scouted for good land in southern Ohio in 1751 on behalf of the Ohio Company.[21] Still later, Tecumseh was especially fond of the Upper Wabash, not least for its corn-growing prowess.[22] And we have seen that Montgomery County, the site of Crawfordsville, was one of Indiana's two earliest corn-belt counties.[23] Indiana's love affair with corn would in fact become the dominant theme in the state's agricultural history. By 1909, well over one-fifth of the land surface of Indiana would consist of cornfields.[24]

Chapter Six

Pioneering in Central Indiana

As social comforts are less under the protection of the laws here . . . friendship and good neighborhood are more valued. There is more genuine kindness and politeness among these backwoodsmen, than among any set of people I have yet seen in America.

—English traveler Elias P. Fordham, ca. 1820, quoted in Jakle, *Images of the Ohio Valley*, p. 116

Luxury has not yet taken the upper hand here, as in other states of the Union. Drunkenness is rare. The old respectability, the hospitality of the woodsman living alone has remained with the resident of Indiana.

—Traugott Bromme, "The State of Indiana" (1848), p. 141

By the time Indiana became a state in the year 1816, both the Whitewater Valley along its southeastern boundary with Ohio, and the Lower Wabash Valley along its southwestern boundary with Illinois Territory, were growing quite settled. Euro-American settlement in the *Lower* Whitewater Valley had started in the 1780s with Virginians and Pennsylvanians squatting on land which, until the

1795 Treaty of Greenville, the United States considered to be Indian territory. From there the Whitewater Valley stretched northward along Indiana's eastern boundary past Richmond, the seat of Wayne County, and somewhat past the northern edge of that county. Richmond was founded in 1806 by Quakers from the piedmont of North Carolina.

After the War of 1812, considerably more settlers flocked into the Whitewater Valley and also to the west into the less-crowded Lower Wabash Valley. In the latter, government land sales jumped 425 percent from 1815 to 1816, and in 1817 the government land office at Vincennes led the nation by selling almost 300,000 acres. Meanwhile, sales of southeastern Indiana land were almost equally high that year at the Jeffersonville land office.[1]

To the north, central Indiana remained a thick forest until the United States claimed possession of it through the New Purchase of 1818. Even then, Euro-American settlers were slow to arrive. Although the federal government opened new land offices in 1819 at Brookville in the east and Terre Haute in the west, very few settlers had yet entered the New Purchase by 1820 when the state legislature decided to put the state capital there.

That geographical center of the new state was home to over one thousand Delaware and other Indians who lived in villages bordering the West Fork of White River. At the turn of the nineteenth century, Moravian missionaries had tried to steer those Delaware in the direction of Euro-American agricultural practices, including animal husbandry, but they had failed to convince very many Delaware to abandon their own farming customs.[2] Moravian records assert that the White River Delaware were "offered more than once all sorts of farm implements like plows, oxen, etc. so that they might live like civilized people." A Quaker missionary who visited White River from a similar mission that was aimed at the Miami of the Fort Wayne area likewise urged the Delaware to adopt whites' farming methods. He was bluntly countered by an Indian who said, "This man tires his horse in vain. He would better stay at home and work for himself as he pleases. We do not need anyone to teach us how to work. If we want to work we know how to do it according to our own way and how it pleases us."[3]

As of 1820 when state commissioners arrived to designate exactly were the state capital should sit, central Indiana's leading Euro-American resident was William Conner, whose father had worked among Shawnee and Delaware in Ohio and whose mother was a white captive who had been raised by Indians. Along with his brother John, William Conner had been raised among Delaware. In 1801–02 he had joined the White River Delaware, married a Delaware woman, and then set up a trading post on the east bank of White River four

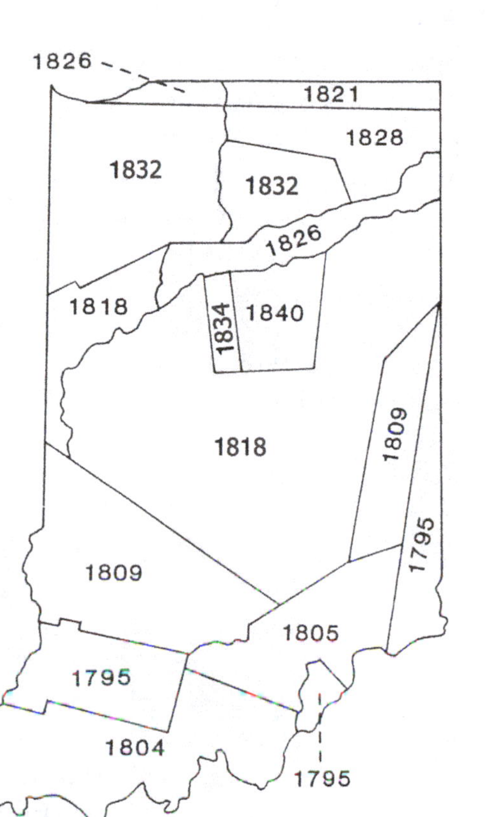

Dates of Indian Land Cessions in Indiana

miles south of present-day Noblesville. Later he played a key role in convincing leaders of the White River Delaware to sign the 1818 New Purchase, and in the summer of 1820 he helped them (including his wife and children) depart for their new homes west of the Mississippi River.[4]

One child who grew up nearby recollected later that, in addition to William Conner and his business partner William Marshall, only thirteen white families then lived in the vicinity. They depended on each other, especially when constructing their cabins. The child recalled the attitude as "'Help me and I will help you;' 'Refuse to help me and you are no neighbor, and you might as well leave.'"[5] They lived on corn as well as potatoes and other garden vegetables, and on plentiful wild game and also plentiful pork thanks to the many razorback hogs that ran semi-wild. Early on, they planted small orchards of apple and some other fruit trees but the apple trees required grafting and were slow to bear fruit.[6]

Those earliest Indianapolis settlers occasionally enjoyed bread made with wheat flour that they bought at the Whitewater settlements, but that was expensive. An escaped slave from Kentucky had lived there quite a while with the

Delaware and he helped supply the new settlers with corn. At first they had to eat it as lye-hominy for lack of any way to grind it. But among their number was a Vermont handyman who had brought along his tools. He devised a small gristmill that could grind corn at a rate of two and a half bushels a day. The apparatus was simply lashed to a tree near the bank of White River. No fee was charged but each user had to supply the horsepower to grind his own corn.

Before long, more substantial gristmills were operating. In 1821 a keelboat loaded with corn was floated down White River from Indianapolis, but business remained slow until a few years later when the aftereffects of the financial crash of 1819 finally wore off and when, in 1825, the state capital finally did move to Indianapolis. At that point settlers started arriving in droves.[7]

Ten years later, in 1835, central Indiana acquired an avid agronomist who wrote lengthy reports about the agriculture there—but wrote them in his native German. His name was Jakob Schramm and his reports about Indiana farming were published in Europe. Schramm was a well-to-do young Austrian hops merchant whose father had disinherited him for marrying a young woman of lower social status. Along with him in 1835 to Indiana came his wife and three-year-old daughter, plus his father-in-law. To help handle the expected work load, they also brought two skilled young men and a twelve-year-old servant girl. Five years in advance of emigrating, Schramm had sent ahead to America a less well-off friend whom he entrusted with the equivalent of several thousand dollars for buying land and beginning the farm-making process.

Jakob Schramm's friend fulfilled his mission by buying land eighteen miles east of Indianapolis near Greenfield, in Hancock County. But by the time young Schramm and his entourage reached that spot in the fall of 1835, he was wishing he had invested in Ohio where his ample money could have bought a farm that was already well-established. Now it was too late. Soon, in fact, he decided that he would do most of the farm-making himself rather than have to pay high wages and put up with the "airs" that many laborers (even recent German immigrants) were flaunting in that heyday of Jacksonian democracy. Schramm explained to his European readers that wage workers in Indiana were referred to as "help" so their egalitarian sensibilities would not be offended, in which case they might quit.[8]

By 1836, when Jakob Schramm described his first Indiana year in a long letter that immediately saw print in Europe, some of the transportation links were already extant that would orient Indiana increasingly toward the northeastern United States. Schramm himself had reached central Indiana from the Ohio River by traveling up the first leg of the new Michigan Road. Back in 1827 Congress had authorized that public land should be granted to the state of Indiana to finance a road bisecting the state by starting from its southeastern corner and extending to its northwestern corner. Indiana's legislature had accepted the federal "challenge grant" in 1828 and had started planning the route. Actual road-building started in 1830, costing an average of $381 per mile.[9]

Barely had Schramm reached his farm that fall of 1835 than he continued on further north, again via the Michigan Road all the way to South Bend, intending to buy more land at the new La Porte land office purely as an investment. That did not work out because the La Porte land office was temporarily closed, so he instead bought 1,920 acres for $2,400 ($1.25 an acre) just northwest of Indianapolis in present-day Boone County.

Returning to his new Greenfield farm, Jakob Schramm launched into its prospects with gusto. He decided cattle were not yet a likely money-maker because not many of them could be sold nearby. Hogs, however, seemed likely to yield big profits because Indianapolis had plenty of pork-packers by that time.[10] In earlier days, many farmers had slaughtered their own "surplus" hogs and had barreled the pork themselves and hauled it to a merchant by wagon, or sent it downriver on a flatboat, but such home-packing of pork had dwindled as commercial pork-packers proliferated.[11] Besides raising hogs, Schramm also fancied he could make money breeding horses and maybe sheep. Sheep, however, required lots of care, and the prices paid for wool were unpredictable.[12]

Most of Jakob Schramm's eventual wealth apparently began around the year 1850 when he sold his Boone County investment land for $4 an acre (thereby more than tripling his investment) so he could buy stocks in a railroad that would connect central Indiana with Lake Erie.[13] That paid well, especially when Schramm foresaw the Crash of 1857 and sold off all his railroad stocks just in time, putting the money into Indiana state bonds. Those paid only 6 percent interest but they did not go belly up after the 1857 Crash, as they had after the Crash of 1837.[14]

Jakob Schramm's farm-making labor took a toll on his health. After twenty years of it, and despite buying the Greenfield area's first McCormick wheat reaper in 1853, he had to retire and turn over his farm's management to his two sons. Among other new chores, they then drained and tiled the farm's wetland.[15] Farm work, as they say, is never finished.

The work that proved to be the most onerous, costing the heaviest toll in health, was women's work. Division of labor by sexes was taken for granted. Men did men's work and women did women's work—except that many wives and daughters were expected to help with field work at planting and harvest time. Rarely did men help women with any womanly chores, which embraced all the work associated with the house, garden, and children. So essential was women's work that men could rarely achieve success as farmers without a woman's presence.[16]

Women's chores included gardening and gathering to obtain food, tending chickens, ducks, and geese, preserving and cooking all kinds of food, keeping house, carding wool and flax fiber and spinning wool and linen thread,[17] weaving cloth and then bleaching or dyeing or fulling it and making clothes from it, washing clothes, plucking geese, making featherbeds, sewing quilts and comforters, knitting socks and mittens, making soap and candles, rearing children, and caring for the sick. Women also made butter and cheese from milk, looked after the flock of chickens, and often traded eggs, chickens, butter, cheese, and beeswax with storekeepers for imports like coffee, tea, and salt. By 1860 eggs were fetching as much as ten cents a dozen, at least in wintertime.

One of women's heaviest chores was the washing of clothes. In June 1826 an Ohio pioneer named Anna Briggs Bentley wrote: "yesterday I undertook the washing. I had to bathe my face in vinegar and water after I got all the white clothes hung out and a pair of course linnen trowsers, the towels, tea cloths, &c boil'd, when I was so outdone I was obliged to throw them in the tub and go to bed." Six weeks later, in early August 1826 she wrote: "I have had my hands full this week: poultices to make, 2 a day, medicine every hour, and many other attentions, and to cook for harvesters. Jos is making the hay in Mr. McNeely's meadow next [to] the road, Isaac, a very heavy crop. They are hauling it home. I had thier dinners—that is, 5 of them—to cook and send off by 11 o'clock and to get all Nancy's meals separate from the rest . . ."[18]

Some tasks both sexes were expected to share, such as butchering livestock, making maple sugar and apple butter, and preparing for weddings.[19]

Most pioneer husbands outlived their wives and many outlived two wives. The rigors of childbirth, including the scourge of puerperal fever, accounted for much of that mortality but so did the unrelenting workload women staggered under. As of 1860, for every 100 men in Indiana who were aged 50 or older, there were fewer than 82 women.[20] One historian says, "all too many women lost their bloom with their teens, were tired out and run down by thirty, and old at forty. Tombstones in the churchyards bear testimony that many a wife died young, to be followed by a second who contributed her quota [of children] and labors, and perhaps a third who stood a good chance to outlive the husband."[21]

A leading expert on early Indiana adds that "pioneer men and boys . . . were not inclined to show much deference to [women and girls by] assisting them in miscellaneous household duties in repayment for their considerable miscellaneous help with harvesting and other farm work."[22] Overwork was the lot of both "butternut" women from the Upper South and Yankee women from New England and the Middle States. Heavy work expectations based on sexual division of labor, regardless of the health consequences, was their shared lot as pioneer wives. In the newly settled Rock River Valley of northern Illinois, which was a Yankee domain, the visiting writer Margaret Fuller found in 1843 that women's

> part is the hardest, and they are least fitted for it. The men can find assistance in field labor, and recreation with the gun and fishing-rod. Their bodily strength is greater, and enables them to enjoy both these forms of life. The women can rarely find any aid in domestic labor. All its various and careful tasks must often be performed, sick or well, by the mother and daughters. . . . The wives of the poorer settlers, having more hard work to do [as pioneers] than before, very frequently become slatterns; but the ladies, accustomed to a refined neatness, feel that they cannot degrade themselves by its absence, and struggle under every disadvantage to keep up the necessary routine. . . .[23]

As mentioned, pioneer women and girls were expected to grow the garden, draw the water, cook the food, clean the house, make the clothes (including cleaning and carding the fiber, spinning the thread, and weaving the cloth), mend the clothes, manage the children (including their education), and care for the sick (with herbs, poultices, etc.). After the settlement of an area was well along, a farm wife could sometimes find a neighborhood girl to hire and thus get help with her household chores, but the earliest pioneers generally found no one available to hire—especially no one female. If they did not bring a hired girl along with them (as the Schramm family did) they generally had to do all the chores themselves.[24]

Under those conditions, and with means of sustenance far more plentiful than was the labor to actualize it all, it is hardly surprising that human reproduction flourished. As of 1820 Indiana held 2.24 white children aged nine or under for every white woman between the ages of sixteen and forty-four. As Indiana filled with settlers, however, that figure dropped by 1830 to 2.14 children, and by 1840 to 1.84 "as the frontier passed through and beyond the state."[25]

Pioneer women's lives in many ways resembled Native American women's lives, but in two major respects they differed. First, pioneer women were not in charge of most of their family's farming, whereas Indian women were. European women had lost the leading role in agriculture thousands of years earlier with the introduction of draft animals and heavy plows (if not even earlier). Their loss of agricultural preeminence had lost European women control over the distribution of food—whereas Native American women *did* largely control food distribution.[26]

And the second major difference was that Euro-American pioneer women lived isolated from each other on separate farmsteads, unlike Indian women, who, in their communal summer villages as well as in their smaller winter bands, had each other as steady partners.[27]

It is small wonder, then, that few women relished the idea of moving to a frontier. "It has generally been the choice of the men," Margaret Fuller reported, "and the women follow, as women will, doing their best for affection's sake, but too often in heartsickness and weariness."[28] Heartsickness came from leaving their childhood homes and their relatives, especially their parents and siblings. Sometimes the sadness was softened by family members following later and settling in the same frontier neighborhood. That was common, in fact. As historians Andrew Cayton and Peter Onuf point out, "New Englanders, Germans, and Irish moved in family units, and recent historians have documented the importance of kinship in determining where and how people moved."[29]

Explaining more graphically, another historian points out that "early settlers came to the Ohio Valley in kinship-neighbor groups that migrated cohesively and sequentially to the region over a period of several years, the typical settlement pattern of the trans-Allegeny."[30] Central Indiana, in fact, is one of the places where such family-mediated migration has been meticulously documented.

The label "stem-family" is a loose translation of *famille-souche*, the label the early French sociologist Frederic LePlay applied to the lineal family reinforced by the horizonally extended family, embracing brothers and sisters who have emerged like branches from the same parental stump or stock. In rural France in the mid-1850s, LePlay found a pervasive rural pattern in which the "stump" of a family (the parents) maintained a homestead from which the brothers and sisters of the next generation branched out—but to which they could return as a haven of safety even after they had families of their own. American sociologists have developed this model further upon finding that family members who migrate to the same vicinity often then begin to reproduce the "stump" itself by creating a new safe haven in the new location.[31]

Stem-family migration into central Indiana has been meticulously documented by the historian Stephen A. Vincent, who has put under a microscope the migration of African Americans to the "Beech Settlement" in Rush County and to the "Roberts Settlement" in Hamilton County. Vincent has discovered that most of those migrants left the same original "hearth" and went to the same destination, but that they did so at different times.

Such family-mediated migration was not unique to African Americans, of course,[32] but Vincent's detailed study of it *is* virtually unique. He found that most of the settlers in central Indiana's Beech and Roberts settlements had migrated there from three contiguous counties which lay along the boundary between North Carolina and Virginia, their migration to Indiana stretching from 1829 through the 1850s. The pace of their migration was especially heavy right after Nat Turner's 1831 uprising in Virginia's Southampton County, which lay just to their east.[33]

Stephen A. Vincent's book on the Beech and Roberts settlements is a masterful example of "the new rural history."[34] He draws on many types of documentation to discover the life stories of people whose lives are otherwise virtually inaccessible to historians. He uses information from land deeds, tax records, marriage records, probate records, and from the questionnaires that census workers filled out as they made their rounds every ten years. Family records, including preserved letters, also help Vincent and other new rural historians reconstruct the life stories of ordinary people.

In the case of settlers in the Beech and Roberts neighborhoods, letters reveal a lot. By 1831, the year of Nat Turner's rebellion in southern Virginia, only a few African Americans had yet come to Rush County's Beech neighborhood, but those few were receiving letters from relatives along the North Carolina-Virginia border asking what Indiana was like. One family in North Carolina, for instance, had received a letter from a cousin who was already living in the Beech neighborhood, a letter telling about the land prices there and local conditions. A member of that letter-reading North Carolina family then wrote back north to his own brother, who likewise was at the Beech settlement, and asked for a letter directly from *him*. The cousin's letter, he wrote north, "gave us great satisfaction as respects your country and we should be very glad to receive one from you." Perhaps the family members still in North Carolina were angling to hear directly from their brother about how much *help* he would give them if they came north to Indiana.

The reply the brother sent back to North Carolina mentioned everything from weather conditions and consequent harvest expectations to the prices being paid locally for crops, and on to the prolific rates of hog reproduction and corn production. "If you could be here," wrote that Indiana brother, "I could go with you in some fields that would make you open your eyes." But in the letters that survive he did not explicitly say how much help he could give members of his family if they did indeed come north.

The Indiana cousin who had first written then added his own two cents to this new letter: "I may inform you all that I have lived this year like I want to live. As to eating, . . . I have had aplenty and some to spare and I expect to make aplenty this year for I have got aplenty of hogs now." As more of the North Carolinians inched toward leaving the South, their questions grew more detailed, asking even about specific land parcels that might be for sale. In the end, family members who were already settled in Indiana often personally went back south to accompany their relatives who finally decided to move north.[35]

As of 1840, the Beech neighborhood held 399 African Americans. But the best land in that area had been bought up by Quakers back in 1820 as soon as the government had offered it for sale.[36] By 1834 the African-American neighborhood was expanding rapidly and *all* the government's public land there, even the worst, had been purchased. Indeed, out-migration of African Americans from the Beech neighborhood began well before the in-migration had ceased, and during the 1840s the number of African Americans there fell from 399 to 349. Beech out-migrants helped to create African-American neighborhoods in the Lost Creek area near Terre Haute, and in Michigan's Cass and Van Buren counties. Most particularly, Beech out-migrants soon created the "Roberts Settlement" in a remote part of Hamilton County, the county just north of Indianapolis whose county seat was Noblesville.[37]

Later, another departure further northward took place soon after Congress passed the harsh Fugitive Slave Law of 1850. When that law passed, many of Indiana's African Americans moved north to Michigan's Cass County, which lies just beyond the two states' boundary. Others moved even further, some as far as Ontario.[38]

During the mid-1830s, the economy of the United States experienced the greatest boom in its history up to then. Farm families shared in the profits by growing more "surplus" crops and raising more "surplus" livestock. ("Surplus" was what early farmers called their output beyond their own home needs.) During that 1830s decade of economic expansion, farmers not just in Indiana but throughout the United States started acquiring more labor-saving devices, and also better seeds, to help them expand their output.[39] And meanwhile, Indiana's state government earmarked ten million nonexistent dollars for transportation enhancement so that more farm products could be shipped out-of-state.

When the 1837 financial panic torpedoed that boom, Indiana's economy staggered. Some of its banks went bankrupt and the state government defaulted on its loans. But its main transportation projects continued (as we will see in the next chapter) and the "surplus" output of Indiana farms continued to grow. For central Indiana, farming's commercialization was hastened just two years after the 1837 Crash by the erection of a large flour mill on the Ohio River at Lawrenceburg in the southeastern corner of the state. Lawrenceburg immediately became a magnet for wagonloads of wheat not just from southeastern Indiana but from central Indiana. A wagon pulled by a single team of oxen or horses could transport seven to nine hundred pounds of wheat. Thousands of such wagons made the trek from central Indiana to Lawrenceburg, where the new mill gladly paid for wheat "in-kind" by dispensing coffee, tea, salt, iron products, dyes, and other assorted necessities and luxuries. Much of the flour ground at the mill was shipped downriver to New Orleans.[40]

After the 1837 Crash, the overall American economy was not very strong again until the mid-1840s, and retrenchment did occur in Indiana as elsewhere. But the state's inherent assets for agriculture were so rich that its population expanded 44 percent in the 1840s despite that decade's relative economic malaise.[41] And even then the top had yet to be reached.

Chapter Seven

Pioneering in Western Indiana

THE GREAT OBJECT is to have as many acres as possible cleared, plowed, set, sown, planted, and managed by as few hands as possible; there being little capital and therefore little or none to spare for hired labor. Instead of five acres well-managed, they must have twenty acres badly managed. It is not how much corn can be raised on an acre, but how much from one hand or man, the land being nothing in comparison with labor.

—English traveler William Faux, 1823;
quoted in Jakle, *Images of the Ohio Valley*, p. 106

As in central Indiana, so too in the Upper Wabash Valley, Euro-Americans began arriving after the vast 1818 "New Purchase." But here too they trickled in only slowly. As of the year 1821, only about 1,500 Euro-Americans yet lived in the New Purchase. But by the mid-1820s the American economy was finally recovering from the 1819 financial crash and settlers were arriving in droves.[1]

At first, most came north by boat on the Wabash River itself, so a federal land office was established in 1819 at Terre Haute to handle Upper Wabash Valley land sales. When more settlers began coming from southern Indiana overland, a land office was opened in 1823 at Crawfordsville to accommodate those migrants.[2] Within a few years, the Terre Haute land office was closed. Crawfordsville also proved a more convenient location for migrants who came straight across west from the state of Ohio. Some of the pioneers realized that their chances of success were aided by choosing a new home where the soil, the terrain, and even the amount of daylight resembled their previous home. Their prospects were brightest where the conditions resembled what their seeds and livestock had previously adapted to, all other things being equal, of course.[3]

Although pioneer settlers came later to the Upper Wabash and points north than they came to southern Indiana, the total cumulative revenue from all the government's land sales there had outstripped each of the state's other land districts by the year 1836. By then, the pace of Upper Wabash settlement was almost frenzied and Crawfordsville's land office was the busiest one in the United States. During 1836 alone it took in at least $1.5 million.[4]

In November 1835 the well-off newcomer from Austria, Jakob Schramm, described the land boom in progress between Indianapolis and South Bend along the new Michigan Road:

> When the land has been surveyed, it is described in the newspapers and notice given that on such and such a day certain lands will be sold to the highest bidder. Hundreds of people have already been roving through the unsurveyed country, choosing what they want to buy. If the land is well situated, for example, along the highway, or on streams where mills can be built, or if it is prairie, an acre sells at from $5 to as much as $20, especially where the prairie adjoins woodland, so that the cleared land can be fenced and there is no lack of firewood. If the prairie is too big, however, and has no woods near, it is not habitable, and is not purchased even if the soil is of the finest. The best and finest of the surveyed lands go first, and at a fairly high price to the highest bidders; later the less desirable lands are sold at the land office at $1.25 per acre. Although perhaps a third of the lands is not sold, everything is gone that is particularly good or well situated.[5]

All that hyperactivity accompanied boom times, the best times economically that the U.S. had yet known. Concurrently in the Fort Wayne area and also in southern Michigan a similar swarm of settlers was arriving from New England and New York State. Sales of public land there were handled by the Fort Wayne land office, which had opened in 1822. That office took in almost as much money in 1836 (at the crest of the economic boom) as did the Crawfordsville office. The new La Porte land office in the far northwest was also quite busy but took in much less money.[6]

Where did all that money come from? Banks created it. Banks at the time were authorized to issue new paper money in the process of extending loans. (That is still true today, but now most loans are issued in the form of demand deposits—that is, as checking accounts.) In 1832 there was only $59 million cash in circulation in the United States but by 1836 there was $140 million in circulation. Until that time, the government's nationwide sales of public land had averaged less than four million acres per decade, but during the three years from 1835 through 1837 the government sold 38 million acres.[7] In Indiana during those three years, the government sold almost 5.7 million acres of public land—over three million of them in the year 1836 alone. Loans outstanding from Indiana banks increased from $1.8 million in 1835 to almost $3.3 million in 1837.[8]

Historians used to assume that *hard* times motivated the settlement of America's frontiers. That assumption fit with the "safety valve" theory, which pictured the frontier as a safety valve where the eastern U.S. could "let off steam" (in the form of migrants moving west) when not enough jobs existed for everyone who lived in the East.[9] But let's keep in mind that most workers back then were farmers, and that it took money (or other assets) to head west and get settled on a frontier with any poise or comfort. *Good* times gave eastern farm families their best chance to sell their eastern land or other assets for enough money to move west and buy public land. During *hard* times, selling their eastern farm would not bring them as much money. Yet, once they reached the West, the base price that they would have to pay for public land was not affected by whether times were good or bad since it was set by law at $2 an acre (until 1820 anyway, and after that at $1.25 an acre).[10]

So it was mainly good times rather than hard times that prompted pioneering. Hard times found most farm folks hunkered down in place and living more self-sufficiently (than they usually did) off their own produce—which otherwise they might have to sell at a loss. In good times farm produce brought higher prices and farm families could potentially profit by moving to the lush West where they could produce and sell more.[11] Admittedly farmers in Indiana had to sell their products much cheaper than eastern farmers were able to sell theirs. In fact, between 1840 and 1846, Indiana crop and livestock prices were more than one-third lower than New York State prices at that time.[12] But on the other hand, land in Indiana was more fertile. And it was also likely to rise more in value than eastern land since Indiana's rural population was increasing faster than the East's rural population.[13] So Indiana settlers could expect speculative gains to supplement their farming income.

In some cases, nonetheless, people did move to frontier areas because times were hard, especially if a frontier was not far distant. Following the Crash of 1819, three families who lived near Cincinnati decided that the deflated prices for their corn and wheat were just too low and their doctor bills were too high. In the fall of 1822, using the almost impassable wagon trails of that day, they traversed central Indiana (passing through Indianapolis) and settled near the Wabash River in Parke County, where one of their number had spied out land on a prior visit. Their years of scarce provisions ended the next fall (1823) with their first harvest. One of the children recalled later that "the first season we made three hundred and fifty pounds of [maple] sugar and ten gallons of molasses [maple syrup] on the same ground that we were clearing for corn." A few more years did pass, admittedly, before they had anyone to trade their surplus farm products to.[14]

That partnership of three cooperating families thus moved during hard times, but what triggered the real stampede into the Upper Wabash Valley were good times, boom times. By the mid-1830s the pace of bargain-priced Wabash land coming newly under cultivation was phenomenal. People were pouring in from all directions.[15]

The low land costs were not the settlers' *only* expense, of course, but the tools needed for farming were still cheap too. Among the early Indiana account books that have survived is one that belonged to a busy blacksmith in Tippecanoe County during the 1830s. Farm tools and what they cost were his bread and butter. The blacksmith David W. Walton recorded selling no soil-turning plows at all (they presumably exceeded his technical threshold) but to buy his "shovel" or "bull-tongue" plowshares—the minimal type of plow described back in Chapter Four on page 42—cost the purchasers only 87 cents. Walton made a lot more ginseng hoes, which were short-handled hoes for digging up ginseng roots, than he made plowshares. His ginseng hoes sold for 50 cents, or two for 75 cents or three for a dollar. Walton's most expensive farm tool was his branding iron, which sold for $4. That was also his price for "boate gunels." His next most expensive farm tool was a big seller, the land-grubbing mattock at $1.50. To replace broken mattock handles he charged 12½ cents and that too was his fee for sharpening mattocks, shovels, and picks. Saw-sharpening cost three times more, 37½ cents. Shoeing a horse also cost his customers 37½ cents, but first they were out 40 cents as his price for the horseshoes. Nails cost two for a penny but screws ran two for a quarter. Pitchforks also sold for a quarter.[16]

If farmers needed their "swingletree" (singletree) ironed (fitted with metal links) they had to pay 50 cents, and that was also their outlay for having a wheel hooped or purchasing a pick. Another mid-1830s source gives prices for scythes, listing $1 for a large, 50 cents for a medium, and 25 cents for a small scythe.[17]

Blacksmith Walton clearly accepted payments in kind because he dealt as well in farm produce. As of 1835 each bushel of wheat he accepted paid off 75 cents of a customer's bill, and three bushels of corn paid off a dollar. Beef paid a customer's bills at three cents a pound, flour at two cents a pound. Two pounds of wool netted a farmer 80 cents' worth of Walton's goods or services, two hundred pounds of hay 50 cents' worth, and a quart of clover seed 25 cents' worth.[18]

Pioneers were not only paying their bills in-kind with what they grew, but often they could pay their hired hands in-kind as well. If they had to pay in cash, the going rate for work along the Upper Wabash in 1835 was a dollar a day, or $2 a day when the hired hand brought along a team of draft animals.[19]

For decades, early pioneers were able to sell provisions to a stream of latecomers. Food, animal feed, and planting seed were among the array of goods that early pioneers sold to settlers who arrived later. Under those circumstances, many of the expenses that earlier settlers incurred were actually investments. To buy a yearling calf cost only $3, and the only expense of keeping it was $2 worth of corn (about fifteen bushels) to carry it over each winter, supplementing winter's sparse forage. Then when it reached four years old, a steer would be worth $30 to $45, and if it had been trained as a work ox it would bring $40 to $60. (A cow, however, would sell for only $10 to $18.)[20] Those were prices *in situ,* which did not involve buying animals where they were cheaper and driving them to sell where they brought higher prices. Often, young steers from northwestern Indiana were driven to Ohio and sold. They then spent their prime years as work oxen in Ohio, and later some of them were driven to the East Coast and sold for meat.[21]

Hogs sold far cheaper of course than even untrained oxen (steers), and far cheaper even than cows, but on the other hand hogs mature uniquely fast. Whereas cattle and sheep retain only about 11 or 12 percent of the calories they consume, hogs retain up to 35 percent. Furthermore, a pig gives birth less than four months after conception and routinely mothers, each year, two litters comprising eight or more piglets per litter. By contrast, a cow carries her offspring nine months and gives birth to only one or two calves at each freshening. A year's profit from 100 hogs at a farm near Indianapolis in 1845 was expected to be $1,000.[22]

Only during the final month or two before slaughter were hogs given corn to fatten them up. If they were fed corn and then taken off it again, they languished. Even merely reducing a hog's ration of corn harmed its health and its propensity to fatten up again later. Thus when hogs were driven to market in the fall, their corn rations had to be maintained or, preferably, increased along the route.[23] Many farms that stood along droving routes maintained "drove stands" including pens and feeding troughs. During the autumn months they did a lively business as overnight stations for herds of livestock and their drovers.[24]

Newcomers in the Upper Wabash Valley clamored for state and federal subsidies that would jump-start canal-building schemes. At the federal level, Congress had promised in 1827 to give the state of Indiana more than half a million acres of public land if it built a Wabash & Erie canal starting eastward from the mouth of the Tippecanoe River (which was considered the upper limit of seasonal steamboat navigation on the Wabash River) and running 160 miles east alongside first the Wabash River and then along the Maumee River, thereby reaching Lake Erie. Congress had made that land grant conditional on the canal-digging work physically starting within five years and finishing within twenty years.[25] The aim was to float farm produce cheaply to Toledo on Lake Erie, from where it could continue east via the Great Lakes and the Erie Canal (finished in 1825) toward New York City.

Indiana's state government "bit the bullet" before that federal offer expired, and eventually the federal government gave Indiana almost a million and a half acres of public land to subsidize the cost of the Wabash & Erie Canal.[26] But until January 1836, Indiana's "bite" remained tentative. Then the state legislature went whole hog by appropriating over $10 million of (nonexistent) money to build not only the Wabash & Erie but also two other canals and assorted turnpikes. The two other canals were the Whitewater, which had already been started alongside southeastern Indiana's Whitewater River, and a Central Canal, which was supposed to bisect the state from southwest to northeast alongside the White River's West Fork, but remained mostly a pipedream.[27]

When a nationwide financial crash struck in 1837, canal-digging slowed. The projects became big money-losers for the state besides bankrupting some private investors.[28] But canal boats finally started shipping farm produce east from Lafayette in 1841 when the canal reached as far east as Fort Wayne. As crop-export costs then fell, the prices that Lafayette merchants paid farmers for wheat rose steadily upward from the pre-canal rate of 45 cents a bushel. The canal's completion on through to Toledo by early 1843 found Lafayette merchants paying farmers $1 a bushel for wheat. Concurrently, the price of bulky *imports* fell just as much. For instance, the price of salt at Lafayette dropped from $9 a barrel to less than $4.[29]

Most of those transportation savings lined the pockets of farmers, and the bonanza prompted hyper in-migration, land speculation, corn-growing, and livestock-fattening for over fifty miles in both directions from the Upper

Wabash River itself, which thereby soon became the heart of Indiana's portion of the corn belt.[30] An early historian said that

> at the various centers along the route were erected large store-houses for the reception of the agricultural products.... Grain was bought and handled through the channels, shipped as fast as it accumulated until winter froze the canal over; then the merchants bought and stored in the warehouses the accumulating purchases of wheat, corn, oats, and pork which were packed during the winter, all to be shipped from the filled-up storehouses as soon as navigation opened in the spring. This kept the canal boats busy until well up to the new harvest. From remote counties grain was hauled by wagon to the canal. The farmers of Grant, Madison, and Delaware counties found an outlet for their wheat at Wabash after a long wagon haul. Similarly, the entire northeastern part of the state and even the lower counties in Michigan hauled to Ft. Wayne. Old settlers tell of long trains of wagons waiting by the hour at these rising commercial centers for their turns to unload the products of their farms, bound to the eastern markets. Four hundred wagons unloading in Lafayette during a single day of 1844 were counted by one of the pioneers. Another, speaking of the business at Wabash, says it was a common occurrence to see as many as four or five hundred teams in that place in a single day unloading grain to the canal.[31]

The canal's annual tonnage peaked in 1856, with 308,667 tons shipped that year. Thereafter, the railroads ate away at the canal's business, which declined steadily for the next eighteen years until, following the financial crash of 1873, the canal was abandoned in 1874.[32]

Why did farmers reap most of the new profits when midwestern transportation costs went down? That was not just a fluke but instead resulted from flexibilities that were inherent in corn-and-livestock family farming as a way of life. The lion's share of midwestern transportation savings went to farm families because those families sat at the controls of a plethora of economic options. For instance, a field of corn could either be shocked and left in the field to serve as livestock fodder or else the ears of corn could be pulled off the stalks and hauled in and stored in a corn crib. If that latter was done, the harvested corn could still be fed to livestock if the cash-grain price was low at the time. Or the corn could be eaten by the farm family, traded to neighbors, or distilled into whiskey and then sold. Even if the corn was fed to livestock (either in the field or at the barnyard) that still left the option of deemphasizing beef or pork production and moving into dairy production. If cities were too distant to sell raw milk, butter and cheese could be made and sold (indeed, both *were* farm-made and sold in large quantities until the late 1800s). And even if beef or pork did remain the choice, that still left the option of dressing and eating the animal at home or trading it to neighbors rather than selling it for commercial slaughter. So Indiana's corn-livestock producers enjoyed many options.[33]

In addition, prior to the Civil War, three different sets of exporters were competing to buy the products of Indiana farmers—namely, the downriver flatboat captains (the region's original exporters), the canal-shipping merchants (who dated from about 1840), and then also the new railroads that came online rapidly in the 1850s. I will have more to say below in Chapter Eleven about this three-way competition for farmers' products, since it helped farmers to maximize their profits.

As for the additional flexibility that farm families gained through their self-sufficiency, the geographer Carl Sauer once pointed out that midwestern farming "from its beginnings, was based on marketing products, but it also maintained a high measure of self-sufficiency."[34] That fact alone led to an array of economic choices, especially since "a high measure of self-sufficiency" was something accomplished by neighborhoods, not by separate households. Neighborly bartering and borrowing, the daily sharing that John Mack Faragher calls "the borrowing system," maximized the pioneers' flexibility and thereby facilitated local self-sufficiency.[35] Their custom of constantly borrowing-in and lending-out meant that even their "fixed assets" were actually "liquid" ones, minimizing their expenses by maximizing the number of people who owed them favors. One key consequence was that corn-livestock farm families usually did not have to sell their output—not at any specific time, anyway, and not regardless of the market price. Until the 1850s, in fact, landowning midwestern farm families were generally able to maintain their landholdings intact whether or not they marketed most of their produce.[36]

What if creditors were hounding them? For precisely that predicament, starting back in the wake of the 1819 financial crash, settlers had convinced Indiana's state government to pass "stay laws" that suspended the collection of debts. Like the corn belt itself, stay laws were another southern initiative. Not just Indiana but Kentucky, Tennessee, and Illinois were among the states in the vanguard of what became a nationwide homestead-protection movement after the Crash of 1819. Even people who owed the federal government money for their land purchases got debt relief. In Indiana, homestead protection culminated in a clause in the 1851 state constitution that made homestead protection sacrosanct. (An 1852 legislative act then set, however, a mere $300 worth of real or personal property as the amount protected by law from seizure for debt of any kind.)[37]

In any case, an array of economic choices fortified the competitive position of corn-belt farm families when canals, turnpikes, and eventually railroads slashed the cost of exporting farm products out of the Midwest. About ten years *prior* to the construction of very many midwestern railroads, in the early 1840s, the average prices that Indiana farmers received for their main products was only 62 percent of the average prices that Pennsylvania farmers were receiving for the same products.[38] But that began changing fast in the 1850s as railroads crisscrossed Indiana. The 1860 U.S. Census reported that "At Cincinnati in 1848 and 1849, (which was the beginning of the greatest railroad enterprises,) the average price of hogs was $3 per hundred [pounds liveweight]. In 1860 and 1861 it was double that.... The cheap prices of the west have gradually approximated to the high prices of the east, and this is solely in consequence of cheapening the cost of transportation, which inures to the benefit of the farmer."[39]

That 1860 Census Bureau analysis was right about farmers reaping the benefit. The transportation savings *did* inure to the benefit of the farmer—until after the Civil War anyway. Over the entire period from 1816 to 1860, the prices paid in Cincinnati for live hogs made up about 80 percent of the eventual wholesale price of the packed pork and lard.[40] And similarly at Chicago, as of 1860 farmers were receiving 84 percent of the wholesale value of Chicago's nine leading farm commodities.[41] In 1860 the midwestern farmer who sold a given amount of farm produce could thereby buy more than twice the amount of manufactured goods he could have bought by selling that much farm produce back in 1816.[42]

But why? First, why did the price of midwestern farm products go up in the Midwest, where they were being produced, rather than going down in the East where they were being consumed? And also why was it farm families who receive the lion's share of the Midwest's rising profits (rather than midwestern merchants or bankers getting it)? Both questions have the same answer: it was because the Midwest's farm families did not *have to* raise food for commercial markets. "Later in the century," explains an economist, "more dependence on the market for farm machinery would pressure the farmer to produce efficiently, to compete and to accumulate; but, as of 1860, the family farmer was not forced or dictated to by the market."[43]

The 1860 U.S. Census report puts it even more succinctly: "The great bulk of the gain caused by the cheapness of transportation has gone to the producer.... The competition of the consumer for food is greater than that of the producer for price."[44]

Midwestern settlers had paid "peanuts" for some of the best crop land in the world. Thus they had multiple choices.[45] They could have lived bountifully in virtual self-sufficiency, especially with the help of their neighborly reciprocity. That neighborly cooperation helped pioneer farm families "hold their own" whether or not they marketed most of their output. But in addition—built like another story on top of their virtual self-sufficiency—good reasons existed why many of Indiana's early farm families went far *beyond* mere self-sufficiency and raised large crops for market. By doing that, they could live even *better* than they could live by simply practicing neighborly self-sufficiency. Merchants had to pay them enough to motivate them to mass-produce food, and that meant farm families had to be paid much larger profit margins than merchants or bankers were able to keep for themselves. It is small wonder, then, that corn-livestock farming—the corn belt—was hastening westward across the Midwest.

Chapter Eight

Pioneering in Northern Indiana

> "A naturally treeless expanse was new to [the pioneer's] experience. He knew that rich soil produced luxuriant plants, but could not realize that plants also made soil, and that grass did a better job of it than trees."
> —Fred A. Shannon, *The Farmer's Last Frontier*, p. 10

As corn-belt farming hastened westward and transformed places like the Upper Wabash Valley, another "belt" of grain-growing was spreading west as well. That was the wheat belt, spreading west a bit further north—coming from New York State and northern Pennsylvania across northern Ohio and into northern Indiana and southern Michigan. From there it would soon continue on west to northern Illinois and southern Wisconsin, then to Minnesota and eventually to the Dakotas, Montana, and Washington State.

Growing wheat to sell for cash, or having it ground into flour to sell more profitably, had been popular in Indiana since the French days at Vincennes. Like the French, many American settlers grew wheat to sell for cash to supplement the corn that they grew for wintering and fattening livestock and for their own consumption. And Indiana's new settlers ate some of their wheat output too, although probably not as high a proportion of it as the Vincennes French. One source estimates that only half of the wheat grown in Ohio, Indiana, and

Illinois south of the National Road in the year 1849 entered into commercial markets.[1]

Southern Indiana was not, at first, part of the wheat belt—although major wheat output did shift southward in Indiana over time. When the U.S. agricultural census was first launched in 1840, it found Indiana's northernmost tier of counties specializing in wheat the most. La Porte County, near the state's northwestern corner produced as many bushels of wheat as it did bushels of corn—which meant that far more *acres* there were being planted in wheat than in corn. All across the state's northernmost tier of counties (seven counties in all) wheat was conspicuous. That 1840 agricultural census found the number of wheat bushels harvested in 1839 in those seven counties to be only one-third fewer than the number of corn bushels harvested there. By contrast, Tippecanoe County on the Upper Wabash harvested *seven times* fewer bushels of wheat than of corn.[2]

Winter wheat was planted early in the fall, and if the first snow fell before the ground surface had frozen hard, that tended to spare the wheat roots a traumatic winter, so the prospects were good. But then, if the spring months of May or June brought warm rain showers alternating with hot sunshine, "black-stem rust" could wreck the whole region's wheat crop—as happened in 1840.[3] Weevils, midges, chinch bugs, and the Hessian fly were the other leading banes that taxed the patience of wheat growers. Only gradually were ways found to control their ravages.[4]

But at first optimism prevailed. The very first step, of course, was breaking the sod by cutting through its tangle of roots. That took a plow both heavy and sharp. Midsummer was the best season for sod-breaking. The requisite plowshare weighed between 60 and 125 pounds and it took from three to seven pair of oxen to pull the rig and turn over a swarth of sod up to two feet wide. The depth of the cut had to be at least five or six inches (which often it was not) or else the sod that turned over would probably dry out and solidify, and not break down completely for several years, obstructing work in the field.[5]

Breaking plows that would "scour" prairie sod did not exist until the mid-1830s, when John Lane and then John Deere, both of Illinois, figured out how to make them using steel. Those plows stayed quite expensive until Andrew Carnegie started America's cheap-steel industry in 1867. Meanwhile, back in 1857, James Oliver had started solving the prairie plowing problem a different way with chilled iron that scoured prairie sod as well as did John Deere's steel plow. By the early 1870s, James Oliver was mass-producing chilled-iron plows at South Bend, and by 1880 his plow factory there was the largest in the world.[6]

After the first breaking of prairie sod, wheat generally could be sown each fall by merely broadcasting it and harrowing it in. If the breaking plow had cut deeply enough through the sod the first year, then the second fall's sowing could even dispense with harrowing. Wheat production began to be mechanized about 1840, but even before that happened, three men could plant and care for several hundred acres of wheat as easily as those same three men could plant and care for a mere sixty to eighty acres of corn.[7] That still left the problem of *harvesting* wheat. But as mechanical wheat threshers were mass produced in the 1840s, and as efficient mechanical wheat reapers proliferated in the 1850s, the time shrank that it took to harvest wheat. A reaping machine operated by three men and pulled by three or four horses could harvest eight to twelve acres of wheat a day—two or three times as much as the same three men together could harvest the old way with cradle scythes. And by the late 1850s, reapers that then also raked the wheat were saving some farmers even more time.[8]

Cyrus McCormick estimated in 1859 that 73,200 reapers were then in use west of the Allegheny Mountains. The main reason there were not more was because cradle scythes cost on average only $4.23 that year, whereas a reaper then cost about $125 to $150. Then too, cradle scythes could last a lifetime whereas reapers wore out in about five years. If a farm family wanted the self-raking feature on their reaper, that added another $40 to the price. By contrast, the reapers that were designed to do double duty as hay mowers soon became quite enticing to potential purchasers because that feature added only a few dollars to a reaper's price.[9]

Reapers were not commonly used until about 1853–54 when their efficiency improved and, at the same time, wheat prices happened to skyrocket. But a full decade before that happened, mechanical *threshers* were already saving farmers considerable time and effort. In fact, steam-driven threshers were quite common in northern Indiana by 1845 although they were still a novelty in most other parts of the wheat belt.[10]

The traditional way to thresh grain was to lay it out flat and then either pound on it with a flail or else drive farm animals over it (which was called treading it). The best results came from swinging the flail, which was more accurate than treading. Animals' hooves tended to damage not just the grain but the straw. Flailing was far slower than treading but conscientious farm families often liked to do their threshing in small batches anyway. If a flour mill was located nearby, frequent threshing in small batches could provide fresh flour for their own home baking and also provide fresh straw to feed their livestock. (Old straw lacks the nutritional value of freshly threshed straw.)[11]

But that was the old way. And as early as 1841 in northwestern Indiana's La Porte County, mechanical threshers were not only in use but were being put on wheels and pulled through the fields, thereby eliminating the need to haul the wheat sheaves to the barn for threshing there. The wheat sheaves were simply fed into the mobile thresher, which tossed off the straw, blew away the chaff, and poured the grain into a box.[12]

Threshing machine, wagon, and farmers in Indiana in 1912

Much of the threshing crew of eight to fourteen men consisted, customarily, of neighbors. The owner of the threshing machine would bring along only a few skilled workers to handle driving the horses that supplied power to the thresher, and, most crucial of all, to feed the wheat sheaves (top first) into the mouth of the thresher. Just before that crucial step, the twine that held together each sheaf of wheat had to be cut with a specially designed knife. Ideally, two twine-cutting specialists fed opened sheaves to the specialist who was feeding sheaves into the machine—one of the two twine cutters preferably being right-handed and the other left-handed. Skill was also important at the other end of the machine, the receiving end, because how the degrained straw was then stacked determined whether or not it would shed water, and whether or not it would even stay stacked when it was later being disassembled day by day to feed livestock and provide their bedding. Generally the home-farm farmer took charge of stacking, since it was he who would later inherit the consequences.[13]

Until it was time to thresh, said a Lake County writer, every farm handled its own crop, but when the

> step of threshing grain is reached, then the work becomes a community project, just as it has been since pioneer days. Usually a certain section becomes a "run" for the owner of a threshing machine, and in many instances . . . the same threshing crew is serving a second and third generation on the same farm. Starting at some certain farm, the machine moves from farm to farm, until the circuit is completed. From twelve to twenty farmers usually assist each other, depending upon the acreage harvested, distance of fields from the machine, and condition of grain. . . . Years ago grain was usually hauled to the barnyards, where it was stacked and allowed to dry until threshed. . . . The farm women of the community pride themselves upon preparing meals for the threshers which earn them the reputation of good cooks. Here too, community cooperation is shown, as usually, several women living on neighboring farms assist each other in the preparation of meals. On large farms where the threshing job may require several days, baking and cooking food for a large number of hungry men whose appetites have been whetted by vigorous exercise in the fresh air is no small task. The dishwashing is a large task in itself. However hard the work may be, threshing time is usually an enjoyable time, for the social aspects are enjoyed . . .[14]

It is clear that wheat-growing families knew how to minimize their cash outlays by exchanging workdays with each other. A writer in Shelby County in east-central Indiana gives even more details:

> The old time "threshing runs" [were] composed of the farmers in an immediate neighborhood [and they] would meet a few weeks before the grain was to be harvested for the discussion of the storage of the grain, extra help needed, what kind of help and tools each farmer should furnish, the routine of the threshing in general, and the rate of difference to be paid. The women met to discuss plans for the dinners and suppers that would be served at the respective homes on the day of the threshing. All the men the Sat. night before the day fixed to start threshing would go to town and buy a new suit of blue denim overalls and waist. These would be worn the first day of the threshing and were worn the entire time of the threshing season. The women would buy calico and make new dresses for this occasion.[15]

The basic process of machine threshing did not change much from the 1840s to the 1880s, but then it changed a lot because steam-engine threshers (also called locomotive threshers) became widespread. That speeded up threshing so much it was no longer necessary to stack the wheat sheaves before they were threshed—or afterwards either if a blowing machine was used to put the straw into stacks.[16] Steam-engine threshing brought neighborhood cooperation to its highest level because of the teamwork such rapid threshing required. Wheat sheaves—also called bundles, or shocks—were brought by wagons straight from the field to the thresher. One frequent result of the new steam technology, however, was lower-quality wheat because it became easier for bad weather or some other glitch to spoil the outcome.

In southern Illinois as of the 1920s, some farms held such extensive grain fields that a thresher and its crew of neighbors might stay at a single farm for up to two weeks, not just eating there but sleeping over as well.[17]

Finally in the 1930s, says a source from Decatur County, "the extensive use of the combine marks the beginning of a new era in harvesting methods and spells the end of long drawn out threshing runs where entire neighborhoods go together to thresh the grain."[18]

All that cooperation had come west with the wheat belt. Wheat-farming families from New York, Pennsylvania, and Ohio had brought it west with them starting in the 1830s.[19] Nonetheless, the newly arrived wheat farmers in northern Indiana discovered they were acutely vulnerable when the financial crash of 1837 struck, followed by a long depression. After 1841, Lake Michigan shippers' market share of sales to merchants in the already depression-hit eastern markets dwindled because the new Wabash & Erie Canal was by then shipping Upper Wabash wheat and other farm goods toward New York cheaper than Lake Michigan shippers could ship them. By fall 1842, wheat was bringing just 38 cents a bushel in Chicago and farmers in northwestern Indiana were lamenting their overspecialization. They began looking into merino sheep and Berkshire hogs, and even into manufacturing possibilities that could replace imported goods with locally made goods.[20]

An added spur to diversify arrived that winter, in January 1843, when the winter wheat crop froze—although that particular loss was largely recouped by planting spring wheat a few months later in the same fields.[21]

But most northern Indiana farm families *did* indeed escape from their overspecialization in wheat. After the 1849 and 1850 wheat crops failed, Solon Robinson used the pages of the *Prairie Farmer* to point out the advantages of corn combined with livestock. "Is it not time," he rhetorically asked, to start realizing "that wheat is not the most natural and staple crop of this part of Uncle Sam's big pasture? Does any land in the world produce better beef than the prairies of Indiana, Illinois, Wisconsin, and Iowa? Grass, either wild or cultivated, is ever growing luxuriantly upon an inexhaustible soil. Indian corn, the best crop in the world for beef, rarely, if ever fails."[22]

The reason why corn rarely failed was not solely its hardiness vis-à-vis the weather but its relative invulnerability to diseases and pests.[23]

Even though Indiana ranked second among the states in wheat production in 1859 (led only by Illinois), much of the crop by then was being grown in east-central and far southwestern Indiana. After another twenty years, in 1879, the far southwest (the Lower Wabash River valley) held three of the state's four leading wheat counties. And by the century's end in 1899, when Indiana had fallen to seventh rank among all the states in wheat output because of the Great Plains' wheat boom, almost *all* of Indiana's wheat was being grown in the state's east-central and southwestern sections.[24]

In effect, the wheat belt had "gone west." But the corn belt had not, not as far west anyway, because corn needed more rain than wheat. William N. Parker sums it up this way: "The frontier crop—wheat[—was] not replaced by corn-feeder operations and mixed general farming because of a restless urge for novelty. [It was] replaced because [it was] driven out of markets by supplies yet further West."[25]

Let us return to that question of manufacturing jobs replacing family-farm life. Images of Chicago may come to mind, but the site of Chicago did not hold even a dozen cabins until 1830 and it was not bought from the Potawatomis until 1833. Despite a froth of speculative frenzy just before the Crash of 1837, Chicago did not amount to much until almost 1850. In 1847 Horace Greeley said it was "filled with land sharks, outright thieves and blackguards."[26] George Ade's father wrote, "I was a resident of Chicago during the summer of 1849. It was then a pretentious country town."[27] Chicago did pack some meat before then, but as of 1850 it packed fewer than 50,000 hogs a year and it did not have an efficient stockyard until the Union Stock Yard opened on Christmas Day 1865. Yet, from the mid-1850s the population of Chicago had been mushrooming. By 1860 Chicago's population was above 112,000 and by 1862 it was packing more pork than "Porkopolis" (Cincinnati). By 1870 Chicago also constituted the country's largest cattle market.[28]

By that time Chicago was drawing some of Indiana's farm families into its wage jobs.[29] Chicago was also exerting magnetism on Indiana farm life in other ways too, such as by rewarding farmers in far northwestern Indiana who converted to dairy production. Lake County by 1870 had become part of Chicago's "milkshed" and its number of dairy cows was the second-highest among Indiana counties.[30] In dairy innovations, Lake County was second to none. It led the way from 1870 onward as Indiana's overall dairy output "completely changed from farm butter and cheese to whole milk, cream, and butterfat."[31] As of 1870, only 936,903 gallons of Indiana whole milk was yet sold commerically, but that figure soon leapt by multiples, with Lake County's commercial sales leading the way.[32]

But an earlier urban magnet for Indiana's northwestern corner, well before the rise of Chicago, was Indiana's own Michigan City. Sited on the big lake's shoreline in La Porte County, Michigan City was founded in 1833 because Trail Creek emptied there into Lake Michigan and that creek's estuary seemed to be the best potential harbor for the terminus of the state-sponsored Michigan Road, which soon arrived from South Bend. (The direct route from Indianapolis to Lake Michigan had been found impassable due to the vast Kankakee Marsh.)[33] By 1836 Michigan City's population had leapt to almost 3,000 ambitious souls, most of them from New England and New York State.

What that did to nearby land values can be imagined. In 1833 a Dunkard family had bought 800 prairie acres nearby for $1.25 an acre. Soon that family found itself smack in the path of the Michigan Road and only seven miles southeast of Michigan City. By 1836 the Dunkard farmer was being offered $40 an acre, even for parts of the farm never yet plowed. The Dunkard farmer, however, informed all inquirers that he preferred huckleberries to wheat and was not interested in selling any land. (Since he had sixteen children to provide for, perhaps his taste for huckleberries was augmented by a desire to endow his children with farms near a market outlet.)[34]

From 1837 to 1844 Michigan City remained northern Indiana's primary grain market. It received thousands of wheat-filled ox carts every year and launching hundreds of wheat cargos onto Lake Michigan in schooners and small steamboats. Ox-cart caravans of up to thirty carts, each cart pulled by four to six oxen, inched up the execrable Michigan Road sometimes hundreds of miles, bringing backcountry merchants' wheat collections and their other accumulated "in kind" receipts to exchange for wholesale merchandise that they could sell or trade at their stores back home.[35]

By 1844, however, Michigan City's commercial primacy ended, a victim of the Wabash & Erie Canal and the arrival of early railroads. Later came a few years when Chicago's merchants were able to pay more for wheat than Lafayette's canal shippers were paying, but that was again reversed when a railroad reached Lafayette from Indianapolis in 1853,[36] and then one from Toledo in 1855.

Meanwhile farther west, in the vertical tier of four Indiana counties that form part of the vast Grand Prairie (most of which lies yet farther west, in Illinois) a slower story was evolving. Yet this story's roots likewise predated Chicago and its market for farm goods. In the 1820s a new generation of cattle specialists in southern Ohio's Scioto Valley had pondered their next move, and by 1829 some of them had relocated to Indiana's portion of the Grand Prairie. The mid-1830s then saw long cattle drives start emerging from the Grand

Prairie, some going all the way to Philadelphia. (These thousand-mile cattle drives continued until the late 1850s, as we will see below.)

At the height of the 1830s' economic boom, dozens of New York State and New England investors each bought thousands of acres in Indiana's part of the Grand Prairie. One of those investors was the Senate's famous orator Daniel Webster, who hired an on-site range manager and paid him a whopping $2,000 a year. Others among the investors moved out to supervise their new cattle spreads in person and built sumptuous mansions to house themselves in style.[37] Well-to-do southerners likewise invested in Grand Prairie land in Indiana and Illinois during the 1834–1837 economic boom, but few or none of them moved in person to their acquisitions.[38]

The Grand Prairie was better suited for cattle than for hogs. Cattle did not necessarily get fat there, but they liked the high, strong, course marsh grass and a lot of it was still free for the grazing in Lake, Newton, and Benton counties well into the 1870s. The cattle were allowed to enjoy it all spring and then in June the herders would burn it off, partly to kill the flies it harbored. New grass would then come up that could be grazed the rest of the summer and fall.[39]

Until a compelling cattle market developed at Chicago in the mid-1850s, long drives to the East kept emerging from the Grand Prairie. A large-scale entrepreneur named Edward Sumner, whom George Ade later called the reigning "suzerain" of Benton County, would in the late summer and fall have his herdsmen start a drive of 100 cattle eastbound every two weeks, and he himself would be waiting in New York City to market the cattle on arrival. Those and other long cattle drives continued leaving the Grand Prairie until the latter 1850s.

Frequently the plan was to drive the cattle east in two stages—first to get the herd to Ohio, where it might be dispersed, but anyway the cattle would be refattened during one or two summers before being driven on to the East Coast. Without Ohio as a way-station, each member of a herd would lose about one hundred pounds during the long drive east. When a railroad bisected Indiana's corner of the Grand Prairie in the 1850s, that put an end to the long drives and the hundred pounds was saved by shipping the cattle in railroad cars.[40] Although those cattle lost little weight en route, they were denied water during the journey and arrived in New York City "battered, bruised, and greatly dehydrated."[41]

From the 1840s on, some Grand Prairie cattle were also driven to Chicago for slaughter, to be used for fresh or barreled meat. Effective refrigeration was still well in the future, however, and Chicago itself did not consume large quantities of food until its population started soaring in the mid-1850s. Salting or pickling beef and packing it in barrels did not preserve it as well as those steps preserved pork. Civil War soldiers who groused about "embalmed beef" were technically correct. Soon after the war, however, using blocks of ice to turn meat lockers and railroad cars into "ice boxes" became widespread and that practice started to solve the beef problem. People like Gustavus Swift and Philip Armour finally solved it completely in the early 1880s by using chemical refrigeration, but by then the Illinois bulk of the Grand Prairie was being drained and converted from pasture to cropland,[42] finally leaving only the still-sizable Kankakee Marsh to frustrate its would-be improvers.

Chapter Nine

Solon Robinson, Hoosier Agrarian

Tall and lanky in figure, with long hair and beard, deep piercing blue eyes and overhanging brows; at one moment his burning glance carried instant conviction of the flaming spirit within; at another his eyes changed to a cool, quizzical, and humorous aspect, revealing the keen intelligence that governed that spirit. In his youth and early manhood he had red hair. Before he was thirty it turned white.

—Herbert Anthony Kellar,
in Kellar (ed.), *Solon Robinson*, vol. 1, p. 41

If we glance as far northwest as possible, at northwesternmost Indiana, we implausibly find one of America's most enthusiastic agricultural authors, Solon Robinson. His enthusiasm must have been congenital, since he alone seemed to enjoy farming in Lake County. Later, one of that county's agricultural agents put it mildly when he said, "Lake County presents a greater variety of farm problems than probably any other county" in the state, not least because it "has about 24 different types of soil ranging from sand to peat."[1]

Solon Robinson settled there in 1834, before almost anyone else. His well-known writings do mention some of the challenges that Lake County's pioneers faced, but his tone remains upbeat, unlike that of other Lake County farmers. An English family named Woods, for instance, came to Michigan City in 1836 and one year later Mr. Woods filed land claim "No. 620" with the local squatters' union, which Solon Robinson had founded on July 4, 1836. (More about the squatters' union below.)

The Woods family then endured many years of misfortune, as described here by a boy born slightly later into the family, who reveals what Lake County pioneering entailed in their case:

> The first ten years of life in Lake county were full of the fierce ague [malarial fever] and chills for the new settlers. The great swarms of blackbirds who made their headquarters in the marshes of the Calumet and Kankakee rivers would swoop down on the grain fields of the settlers and it was a fight to the death to save any of the crop for the one who had planted it. My father told me about how one year he planted a twelve-acre field of corn and the birds took every bit of it except one small wagon-box full. He said that he laid down at the end of the field in despair and cried. Most of the settlers who made claims in Lake county in the early days gave up in disgust and pulled out after a short time. The only ones who did stay, my father said, were those who were so beggarly poor that they could not find a way to get out. In fact, the first twenty years of life in the region was an actual fight for existence.[2]

"The ague" mentioned here was malarial fever, and a significant number of Hoosiers suffered from it. Everyday life was organized to accommodate people's recurring attacks, especially during the autumn, when the symptoms were at their worst. The attacks started with feelings of languor, with yawning and stretching. Then came cold chills until one's teeth were chattering. After an hour of cold chills, one's body overheated and severe back pains set in, culminating in profuse sweating as the attack ended.[3]

Malaria's link to mosquitoes was not yet known, but everyone except cranks agreed that the malady was somehow linked to swamps and marshes, to any stagnant water. Travelers and new settlers were warned to avoid those places and most tried to do so, cranks again excepted.[4] The antidotes prescribed in northeastern Indiana featured whiskey and herbal bitters. Reputedly they worked best when imbibed together and then were followed by vigorous exercise.[5]

Regarding that squatters' union Solon Robinson founded, it was one of dozens that proliferated all over the Midwest. Much of the future Lake and Porter counties' still-sparse citizenry showed up for the founding meeting on July 4, 1836. The organization proved so popular that land speculators found themselves powerless. When finally in 1839 Indiana's far northwestern corner was opened for auction sale at La Porte, "all bona fide settlers who desired to buy their holdings at the regular price of $1.25 an acre obtained them without competition. Perhaps the attendance of Solon and his cohorts, well armed, helped to ease the situation [*sic*]. This episode was to give Robinson the title 'King of the Squatters.'"[6] As of 1839, when this land auction occurred, $1.25 was *not* actually the "regular" price of land at government auction sales, but only became the regular price after the auction was *past*. But $1.25 did become the de facto auction price when Solon Robinson and his armed fellow squatters intimidated anyone from bidding up that price after the land's actual resident had bid it. No wonder the La Porte land office took in little money.[7]

Squatters' unions—also known as "claims clubs" and "claim associations"—were good news for bona fide settlers. Where they didn't exist, speculators often used money or stealth to acquire land that squatters had improved.[8] Solon Robinson's own land claim in Lake County had been contested by a crafty and persistent speculator from Michigan, an annoyance which had helped prompt him to start the squatters' union.[9] His unlikely national career as an agricultural expert grew from that experience and was symptomatic of a new quest for "improvement" that began to attract some Indiana farmers in the 1830s and soon manifested in the founding of numerous county-level agricultural societies.[10]

Solon Robinson had been born in Connecticut in 1803 and may have first come west as a Yankee peddler. By 1827 he lived in Cincinnati and in 1828 got married. In 1830 he moved down the Ohio River to the vicinity of Madison, Indiana and spent four years there in many roles, including as a rural realtor and auctioneer. By age 31, however, he was plagued by recurrent illness. With an ox-cart and horses, he and his family traveled the new Michigan Road to the far northwestern corner of Indiana, settling in what would soon become Lake County and befriending Indians who lived there. His brother Milo, who suffered from tuberculosis, soon joined him. They grubbed out farms and protected them from speculators (along with the farms of 474 other local settlers) by founding the squatters' union. In partnership with his brother, Robinson soon opened a store, which until 1840 took in large quantities of cranberries and furs from Potawatomi Indians. Almost no currency was ever involved in the store's transactions, even after the last Potawatomis were expelled west to Indian Territory in 1840.[11]

Robinson was elected Lake County's first county clerk. He also served as the postmaster at what later became Crown Point, the county seat, and used his U.S. "franking" privilege to finance his voluminous correspondence with progressive farmers and agricultural journals nationwide. His letters were avidly

Solon Robinson, 1841

sought by agricultural journals for their charismatic writing. Robinson's eventual career as a novelist was prefigured in 1841 by his first novella, a story about pioneers that a Cincinnati newspaper serialized. But most of his writing in the 1840s was nonfictional agricultural advice. He loved especially to describe his own experiments and their results—experiments with razorback hogs, Berkshire hogs, merino sheep, farm architecture, ways to cure beans, grow oyster plants and mammoth sunflowers, plant trees, drain wetlands, curb wheat rust, make fences out of sod and/or out of thorn bushes (this last a disaster, in his opinion), and other fads of the day.[12]

Robinson's Berkshire hogs were apparently Indiana's first. By the time of the Civil War, Poland-China hogs as well as Berkshires and some other breeds had become common on Indiana farms, but such breeds often fell victim to ailments. The easiest way anyone found to keep pedigree hogs healthy was to let them cross-breed occasionally with old-fashioned razorback hogs. That not only improved the next generation's health but doubled the number of piglets in the litters of the next generation.[13]

By 1852 Solon Robinson had moved to New York City to try to start a national agricultural magazine. That plan soon washed out, but he became the agricultural editor of Horace Greeley's *New-York Tribune*. Like other prairie pioneers, Robinson totally failed to predict barbed wire and thought that prairie farms should copy French open-field farming, using ditches to mark the boundaries between farms and allowing livestock to wander at will during the winters.[14] Some other prairie aficionados championed thornbush hedgerows planted along the top of low "sod fences,"[15] but this was a fad that exasperated Robinson. Later—in the 1850s—a vicious thorn plant called the Osage orange became the fencing rage and remained fairly popular until barbed wire arrived in the 1870s.[16] The barbed-wire era led to planting black locust trees in fencerows so the wire could be nailed to their trunks. Many of the fencerows paralleled roads, however, and a century later when the locust trees aged and blew down, they often took down with them the telephone and electric wires that had also by that time been strung along the roads.

Robinson's most popular writings dispensed advice and warnings to prospective west-bound emigrants. After being reprinted by many journals, those essays were gathered together into books. In 1840 Robinson had helped start a Union Agricultural Society based at Chicago. Through that and many other venues—as a writer, a judge at agricultural fairs, and increasingly as a nationwide lecturer—he promoted agricultural education and experimentation. For several years he spearheaded an effort to start a national agricultural society, but that effort led to naught.[17] (The Grange emerged later, in 1867, and became the first successful nationwide farmers' union.)

One myth about the Midwest that Robinson helped dispel was that prairie fires were necessarily dangerous. Few grass fires even ignited rail fences, he reported, and horses could be ridden right over the fires, since only a thin line of grass was significantly aflame at any given time. Only the tall, thick growths of marshes could make prairie fires dangerous—so Solon Robinson claimed.[18]

Open-prairie farming became Robinson's special study. (Not that he personally still farmed much after his other activities multiplied.) His calculations showed start-up costs for prairie farms to be considerably higher than start-up costs for farms in the rest of Indiana. In 1842 he calculated the expenses entailed in creating a 160-acre farm in northwestern Indiana and they totaled $800—including $200 for the land itself, $240 for sod-busting (the first plowing), and $173 for enough rail fencing to subdivide the farm into four plots of forty acres each. Then, if an additional forty-acre woodlot was bought nearby (as was advisable), another $120 would be needed, since wooded spots on Indiana prairie were no longer available from the government, having been snatched up early by land speculators and pioneer settlers—two categories of people who were often identical, incidentally.[19]

Thus, farm-making costs on the prairie would total about $5 an acre if the farm family itself did all the nonspecialized work. Land speculators who were trying to *sell* prairie land, however, could put a different twist on similar facts. A contrary estimate of farm-making costs had appeared in the 1838 book *Valley of the Upper Wabash* written by the Lafayette land speculator Henry William Ellsworth. Ellsworth claimed that it cost a whopping $12 an acre to have timbered land cleared off, even with the stumps and roots left in the ground, but that prairie land could be plowed and made ready for cultivation for just $3 to $9 an acre. Unlike Robinson's writings, however, Ellsworth's message was aimed at investors. He neglected to mention that the $12 per-acre cost of forest clearing could all be done by a farm family itself (with the help of a neighborly log-rolling) whereas a lot of the prairie start-up cost of $3 to $9 per acre had to be paid in cash or in-kind to a sod-breaking contractor.

Note that all those costs were as of the 1840s. When railroads arrived in the 1850s, the cost of everything went up, especially of land—because, as the *Prairie Farmer* announced from Chicago, the completion of railroads to the East Coast instantly added 10 to 25 percent to the value of midwestern grain and livestock.[20] The race was immediately on to produce more wheat by mechanizing more. And requiring even less investment than wheat machinery was the fattening of more livestock. What ballooned the size of Chicago in the 1850s was the completion of railroads linking it to New York City. By 1856 the number of Illinois-fattened cattle for sale in New York City often surpassed the number of Ohio-fattened cattle for sale there, and by 1863 Illinois was sending five times more cattle to New York City than Ohio was sending.[21] In fact, Solon Robinson used his *New-York Tribune* position to publish weekly reports on livestock arrivals, providing detailed statistics. "Robinson's price quotations, statistics of arrival, and statements of the condition of the market were summarized in many farm journals and had much to do with the flow and price of cattle in the New York market."[22]

Back in the year 1823, when a U.S. army expedition under Stephen Long had crossed northwestern Indiana en route to find the headwaters of the Mississippi River, the expedition's journal-keeper had written, "the country is so wet that we scarcely saw an acre of land upon which a settlement could be made."[23] A few years later, as settlers started to trickle in, the dry prairies that lay distant from stands of trees were occupied only hesitantly and wet prairies were hardly occupied at all. Much later, when tardy settlers finally *did* occupy wet prairies, that was partly because other land had grown costly but also because new ditching and tile-laying methods had been devised to make wet prairies drier.[24]

Where water did not just saturate the ground but covered it, the first step was to cut large ditches to make standing water move toward rivers. Beaver Lake, for instance, which dominated northern Newton County, was nowhere deeper than eight feet but it stretched seven miles in one direction and five miles the other way. An effort to drain it started in 1853 by cutting a ditch that moved its water toward the Kankakee River. Later, following the Civil War, larger ditching efforts finally eliminated Beaver Lake "at most seasons of the year," but it still returned every winter and lasted into spring.[25]

The main ditches were scooped out of the muck with special ditching plows two and a half feet wide. George Ade, a native of the region, said normal ditching operations required fifteen yoke of oxen but that "in mushy ground and bad going," especially if the ground was not only soaked but rough, thirty yoke of oxen were hitched together to do the job. West across the state line in Illinois, one large farm used a ditching plowshare measuring eleven feet long and almost three feet wide, mounted on a rig eighteen feet long and pulled by sixty-eight oxen (thirty-four yoke). A crew of eight operated that rig and it dug three and a half ditch miles per day.[26]

Just draining away the standing water made much land cropable, and settlers often found such land highly fertile. Smaller trenching plows were then devised that could drop four feet or more below ground level. Those special plows scooped out trenches in which cylindrical tile pipes, four to six inches in diameter, were laid so that water would drain through them in the direction of gravity flow. The trenches were then filled back up with topsoil mixed with organic matter. Water entered the pipes at their joints from the lower side.

Those tile pipes were laid in parallel rows about forty feet apart where the subsoil was clay, but up to sixty or even one hundred feet apart where the subsoil was loam. Extremely wet fields could require that the pipes be laid closer together than forty feet.[27] Among the crops that later thrived in the "muck" soils were peppermint, spearmint, onions, cabbage, and celery. Fields of mint were cut like hayfields and two crops each year were the norm. The fresher from the field the better, the mint was distilled in tubs and vats to produce mint oil for use in confections. Then the spent "hay" was spread back on the fields to replenish the soil, a practice that would have warmed the cockles of Solon Robinson's heart.[28]

Perhaps the most poignant of the changes that mechanized farming entailed was the passing of the faithful ox team, superannuated by the speed requirements of the new farm machinery. One writer explained that oxen "will endure more fatigue, draw more steadily and surely; are purchased for a smaller price; are kept at less expense; are free from disease; suffer less from laboring on rough grounds; and perform the labour better; and, when by age or accident they become unfit for labour, they are converted into beef. The only advantage of employing horses instead of oxen, is derived from their speed."[29] Horses walked twice as fast as oxen and many of the new farm machines required horse-speed to work well, since their gears were turned by the axles between their wheels. As farmers mechanized their operations, they retired their work-trained oxen and used horses. The 1850s saw Indiana's farm-horse numbers rise 66 percent. And when the Civil War then brought exceptionally high beef prices, the fate of the faithful ox team was doubly sealed. From 1860 to 1870 the number of work-trained oxen on Indiana farms fell 88 percent, down to a mere 14,088 oxen.[30]

The passing of the faithful team of oxen was but one of many changes that mechanized "railroad farming" fostered. The next chapter glances at several others.

Chapter Ten

The Competition

The effect of new transportation facilities in creating a broad regional market was to bring farmers into competition with each other over widening areas. . . . The pressures were eventually felt by all.

—Clarence H. Danhof, *Change in Agriculture*, p. 15

Farmers had crossed an invisible threshold . . . from an era of making a place to the era of finding a place within a larger whole. As they would find out, it was a mixed blessing.

—Susan Sessions Rugh, *Our Common Country*, p. 125

We have seen the attraction exerted in the pioneer era by easy-entry farming. And now we have begun discussing why, after the Civil War, many small-scale landowners and tenant families had to pack up and move on, leaving their rural Indiana homes to seek farms further west or to begin a town or city life.

Even before the Civil War, grown-up children had often faced the predicament of inheriting land parcels too small to provide a livelihood. In one central Indiana case, the 340-acre farm of an African-American patriarch at Beech Settlement (in Rush County) provided each of that man's twelve children only twelve to twenty-one acres apiece when the farm was unencumbered after his death and then divided.[1] After the Civil War, that kind of predicament was all the more common.

But let us not skip the Civil War itself. In our mind's eye we might visualize vigorous Indiana farm wives taking over their soldier husbands' work roles by hitching the horses to the new machinery and bringing in the wheat and hay. State agricultural officials even used the pages of the *Prairie Farmer* to publish a plea to farmers asking them to let their daughters help do the farm work. The *Prairie Farmer* also carried a large advertisement showing a hay rake being driven by a young woman.[2]

And some farm wives and daughters surely did that, but not most. Women remained exceptional on those driver's seats despite the war. "It appears to have been rare for a woman to do male tasks on the farm or leave their gender defined economic roles," says one scholar of west-central Indiana.[3] He adds that only about one-fourth of the married men in west-central Indiana saw Civil War service anyway, whereas half of its unmarried men did so. Few of either group served through the whole war, some serving as briefly as three months, the equivalent of merely a prewar absence to float farm goods to New Orleans. Others spent one, two, or three years in the service, but even such long absences were not uncommon before the war. Many Indiana farmers had been away that long prospecting for gold in California.

The ways farm families managed to take up the slack were not new either. Farm improvements were postponed if they were not urgent—including the unpopular chore of manure-spreading—and neighbors and relatives were counted on to provide extra help when needed. Boys were often expected during the war to perform more farm work than usual. Some boys thereby earned significant wages. Other less fortunate boys were simply forced to work harder. One boy named John Cutler, whose family had recently arrived from Ohio and had settled in a marshy part of Noble County in the state's northeast, said later that when the war started "they sold me to my uncle," who lived back in Ohio. After a while he ran away from his uncle and joined the First U.S. Cavalry, becoming on one occasion a bodyguard for President Lincoln.[4]

In west-central Indiana and elsewhere, farm families not only "made do" but donated farm produce in large amounts to impoverished urban families which had breadwinners in uniform. Once in 1864 "some fifty-two loaded wagons paraded through the streets of Crawfordsville" filled with food and firewood for poor soldiers' families.[5] West-central Indiana was still almost 91 percent rural then, which helped the countryside to "carry" the cities. Despite the presence of Terre Haute, Lafayette, Greencastle, and Crawfordsville, west-central Indiana was still only 9.1 percent urban, and Indiana overall was only 8.8 percent urban at that time.[6]

However Civil War-era Hoosiers managed to do it, statistics show that when the war boosted demand for food and the prices of food went up, the output of Indiana's farms increased. As the price of corn rose from 26 cents a bushel in 1861 to 79 cents a bushel in 1864, Indiana farms' output of corn went up from less than 72 million bushels in 1859 to over 116 million bushels in 1865, and on to almost 128 million bushels in 1866 after the surviving soldiers were back home. As for wheat, its price rose from 83 cents a bushel in 1861 to $1.51 by 1864, on up to $2.22 a bushel by 1867, and down only to $1.88 in 1868. Meanwhile, Indiana farms increased their wheat output from less than 17 million bushels in 1859 to over 22 million bushels in 1864, and produced over 20 million bushels in both 1869 and 1870.[7]

The high Civil War-era wheat prices followed almost on the heels of the high 1850s wheat prices, which had inspired thousands of farm families to plant more wheat and to buy their first mechanical reapers. Between 1850 and 1854 the price of wheat had risen 60 percent (even when adjusted for inflation) and it had stayed high until 1858.[8] Although the price of wheat then slumped, it soon went back up and soared higher than ever during the Civil War. Those high prices prompted farmers in the five states of the Old Northwest, plus Iowa, to add more than two and a half million acres to their grain crops between 1862 and 1864.[9]

And thousands of farm families bought reapers in the 1850s and particularly in the 1860s. If they had continued using scythes, even using cradle scythes combined with a division of labor, a team of men could only cut, bind, and shock about one acre a day per man. That was a tight constraint, since wheat released its grains within, at most, ten days after they ripened.[10] Until the reaper—and really until truly efficient reapers went on sale in 1854—large wheat fields remained uncommon because they required veritable armies of harvest workers, whether family members, farm hands, or (as earlier at New Harmony) Rappist communards. Even medium-sized wheat fields had often required that scything the wheat begin before its grains were fully ripe.[11]

Wheat's brief two-to-ten-day harvest "window" also convinced most families who grew a significant amount of wheat to buy their own reaper rather than to invest in joint ownership with neighbors or kin, and rather than hiring a contractor who would bring along his own reaper.[12] Wheat-growing was already a gamble without adding additional uncertainty about whether the crucial equipment would be available when needed. Harvest time would not wait, and the machines (which cost $100 to $120) could be bought by substantial farmers on credit pending a harvest. If a farm held plenty of good wheat land, a new reaper could plausibly be paid off with a single harvest's profits—assuming that wheat prices were high at the time and the bushels per acre were high that year too.[13] To qualify to buy a reaper on credit, a family had to farm enough acreage to make the implement dealer feel he was likely to get his money later, or to get his bushels of wheat if the reaper was to be paid for "in-kind."[14]

But the threshing *could* wait, so most farmers patiently waited their farm's turn in the threshing "run," as described above in Chapter Eight.[15]

Not just high prices but something additional in the 1850s was pointing farm families toward wheat. Much of the Midwest's soil fertility was being lost, and many fields would no longer grow much corn. Wheat-growing (and hay-growing) were on the rise not only because their prices were high and because reapers, mowers, and rakes were now more efficient. Those were indeed the "carrots," but the "stick" was that many fields' initial fertility had disappeared through decades of growing corn without manure. Concerned about this, Indiana's State Board of Agriculture conducted a survey in 1857 about manure usage, but the survey evoked little response. Those who answered indicated that nonuse of manure still predominated but that its use was increasing.[16]

Rather than trying to keep fields fertile enough to grow corn by hauling and spreading manure, which entailed heavy work, or by using guano or other fertilizers which cost money, many farmers simply began rotating their crops. Manuring and rotating *both*, of course, would have been best, but manuring was hard work and rotating crops was easy. A standard rotation was to plant a field in wheat for a year or two and then in clover for a year or maybe two, and then to return that field to corn-growing for two or three years before rotating again.[17] So not just high wheat prices and better reapers were motivating the trend toward more wheat, but so was soil that excessive corn-cropping had depleted.

Many other farmers escaped worn-out soil by simply moving further west, and then with the same motive they often moved west again. Most American farmers, says one historian, "failed to internalize the cost of ecological damage even when it resulted in smaller and poorer harvests, and rather than make good what they had begun, they could [move west] to the frontier."[18] When the semi-arid plains came into view, however, such avoidance behavior declined. In Missouri's part of the corn belt, for instance, farmers who had the option to diversify by adding wheat to their crops were almost all still rejecting that option as of 1860—but soon that changed, and even though "the cultivation of small grains [in the corn belt] requires a large sacrifice of corn, hog, and cattle production," nonetheless in Missouri by 1880, just as in Indiana earlier,

"Corn Belt farmers . . . found it advantageous to grow more small grains" and thus the "Corn Belt crop mix became more *diversified*."[19]

As it turned out, the mechanization and expansion of wheat-growing marked a turning point in the American economy. The building of railroads had already become a significant economic "multiplier" by the 1850s, but, by itself, railroad-building had failed to jump-start mass industrialization. That took both the building of railroads *and* an explosive growth of farm-implement manufacturing.[20] It was those two metal-heavy enterprises together that inspired Andrew Carnegie and others to "bite the bullet" and invest the considerable start-up funds required to make cheap Bessemer-process steel in America.

Large-scale metal demands had admittedly first come from railroad builders. American iron output was already going 17 percent toward making railroad rails by the early 1850s. A far smaller percentage of the nation's iron was then yet devoted to making farm machinery. But soon the Civil War prompted a vast increase in farm machinery. In 1858 less than 100,000 reapers and mowers had been in use in the northern United States, but by 1864 more than 250,000 were in use. And by 1870, over one-fourth of U.S. machine-making by value would be devoted to making farm machines.[21] Beyond that, many of the metal products used to refine farm commodities after they left the farm were also being mass produced by that time. Together, farm implements and food-processing machines played a major role in bolstering the U.S. steel industry, not only by the amount of steel used to make them but also by the specialized grades and types of steel they fostered. One scholar notes that "virtually all of the cities that became major steel producing centers, including Pittsburgh, Buffalo and Chicago, had as part of their early industrial bases a large farm implements industry which had exerted local demand for specialty products."[22]

Worldwide by 1900, more money was invested in harvesting machines than in any other type of machine except steam engines, and the value of new farm machines being manufactured in the U.S. had grown beyond $100 million a year.[23] But all this mechanization had a downside. As one historian explains: "The new improved implements had made some of the farmer's hardest tasks easier, had made him less dependent on hired labor, and had enabled him to manage larger acreages in grain and hay, but they had also increased the capital cost of farm making. It was no longer feasible for a person without capital to create a farm on the prairie, and it was becoming more difficult for a man to rise from laborer through tenant to owner."[24]

Here the reader might think I am about to claim that mechanization doomed Indiana's family farms; but it was not that simple, because there was not enough mechanization of *corn*-growing to threaten family farming as yet in the corn belt. Granted that more and more farm output, even of corn, was going to market rather than being used at home.[25] But making that trend possible was precisely the fact that *families* provided most of the labor on farms. Without their numerous children, farm families would have had to hire more farm hands. In that case, not much of a trend toward commercial farming could have occurred, especially since corn-growing was still quite labor-intensive. Hiring farm hands tended to be far more expensive than using the labor of one's own children, who usually cost their parents just room and board.[26] Most hired hands likewise received room and board and beyond that they had to be paid wages, which by 1870 had leapt up to almost $20 a month on average—about 40 percent more than farm hands had been paid on the eve of the Civil War.[27] So even though mechanization cut the average work time per *wheat* acre from sixty-one hours in 1830 to just four or five hours by 1896, it still usually took large families to do everything else that needed doing.[28] As Susan Sessions Rugh found when she studied family farming in west-central Illinois in the 1800s, "The work of all family members was vital to powering the move toward market engagement in the post-frontier phase of agrarian capitalism."[29] Indeed, another historian says that "*community* did not 'break down' with the approach of the modern world; community, in fact, provided a means of making a transition to it."[30]

Meanwhile, the growing of other crops was streamlined far less than the growing of wheat. Machines reduced oats' work time only half as much as they reduced wheat's work time. And the time spent on hay, even hay that was left unbaled, fell only a third as much as the time spent on wheat—from twenty-one hours per hay acre in 1830 down to four hours in 1896.[31]

Let us not misinterpret mechanization as a sign of crass materialism. Although high wheat prices and drastic work-time cuts did inspire the 1850s and '60s leaps in Indiana's wheat output (and would inspire another leap in the late 1870s), that does not mean greed was necessarily the motive behind Hoosier farmers' crop choices. Most farm families *needed* more cash simply to keep farming. Their cash expenses were rising—partly to pay for the new machines but also to pay their rising mortgages and taxes. Mortgage costs were rising in tandem with land prices, and Indiana's tax rates climbed as well. In fact, Indiana's tax rates had already multiplied by eight times between the years 1834 and 1841 while Whigs were running the state.[32] One historian goes so far as to say that "heavy tax burdens forced farm practices which depleted the soil, produced erosion, and diminished land values."[33]

Then too, if a family wished to reproduce its way of life in the next generation, it typically had to buy additional land so that its children could inherit enough land to make viable farms. We have seen land prices soaring already in central Indiana by the 1850s, sending some farm families packing then. Most families did still survive on the land, but often they did so by abandoning equal land division among all heirs (partible inheritance) and adopting, to some degree, "preferential" inheritance—either primogeniture or a variant thereof. Many families compromised so that one son, or perhaps a son-in-law, would receive the family farm intact and then had to buy out the shares of his siblings (or his brothers- and sisters-in-law) through scheduled payments spread out over many future years.[34]

Most farm families were not trying to maximize their profits but instead were trying to sustain and strengthen their position as farm owners. In other words, they were basically trying to avert risks rather than to maximize profits—although, if all *other* things were equal, high profits did help them avert risks.[35]

The 1850s became a window of opportunity for farm families to get themselves securely rooted as farm owners. Indeed, that opportunity further widened in 1854 when Congress suddenly passed a Graduation Act that cut the public-land price to a mere 12½ cents for every acre that had been available for sale for thirty years or more. The price tag on other unsold public land was also reduced below the standard rate of $1.25 per acre, using a formula based on how long each piece of land had been up for sale. Thus after thirty years on the market, public land now cost 12½ cents an acre; after twenty-five years it cost 25 cents an acre; after twenty years it cost 50 cents an acre; after fifteen years, 75 cents; and after ten years, the price was now one dollar an acre.[36] That temporarily took some of the pressure off land prices, for Indiana had still held about three million acres of unsold public land as of 1850, most of it in southern and northwestern Indiana. A million of those northwestern acres were in the meanwhile sold cheaply as "swampland" in 1852–1853, and then almost all the rest of Indiana's public land was sold in 1854–1855 under the Graduation Act. All the land went fast, for neither the federal "swamplands" legislation of 1851 nor the 1854 Graduation Act set any limit on how *much* land could be bought at the cut-rate prices, so most of the land buying was speculative.[37] In turn, that prolonged the wasteful farming methods which cheap land had underwritten all along.[38] And meanwhile, Indiana's population grew a whopping 37 percent in the 1850s.[39]

Yet family farming survived and flourished, thanks mainly to the corn-hog system. Regardless of wheat's high prices and drastic labor savings, few Indiana farm families tried to "maximize" their cash incomes by taking the gamble of growing wheat alone. All through Indiana's bonanza wheat years, the state's overall acreage planted in wheat never came close to its acreage in corn.[40]

In effect, family farm life was being prolonged by inventors' failure to devise an efficient corn picker. That failure is what kept family-farm life going strong through the late 1800s and into the 1900s. Before 1900, machines managed to reduce corn's labor demands by only about three-fifths, from 38.75 hours an acre in the year 1830 down to 15.15 hours in 1896. That saving of labor was achieved by improving plows, seed-corn planters (also called corn "drills"), cultivators, "sleds," "binders," and by efficient "shellers" able to shell many hundreds of bushels of corn a day. Then too, silos for storing cornstalks and leaves as silage were becoming somewhat common by 1900. But what still mattered most was the lack of any truly effective corn-*picking* machine. Corn pickers did start improving somewhat in the 1890s, but they were not very widely used until World War I.[41] Many farms did not acquire one until after World War II.

Having myself once gathered shellcorn ears all day for $10 in a field thicker with cockleburs than cornstalks, I will not rhapsodize here. Yet the fact remains that the eventual invention of efficient corn harvesters did herald the virtual end of Indiana's family farms. Despite all the new farm machines that had already appeared and improved between 1850 and 1890, the percentage of Indiana's male work force engaged full time in farming fell only from 66 percent in 1850 to 59 percent in 1890. Then followed the major era of Indiana's industrialization, from 1890 to 1920, and that brought the percentage of Indiana's work force engaged full time in farming down to 31 percent by 1920.[42] The widespread adoption of mechanical corn pickers during World War I had some effect too, but those machines did not actually work very well until they were driven by tractors rather than by horses, and that changeover was not very common until the mid-1920s. As of 1920, only 4 percent of Indiana farms as yet had a tractor, but by 1930 that would rise to 22 percent, by 1940 to 37 percent, and by 1950 to almost half the state's farms.[43] Meanwhile during those same decades, not just mechanical corn picking caused enormous labor savings but also, starting in the 1930s, the mechanical chopping-up of corn plants to make silage. Then in the wake of World War II, herbicides rapidly replaced mechanical cultivating as a far faster way of controlling weeds. The combined effect of these innovations has been called "the corn revolution."[44]

In fact, not just tractors, corn-choppers, and herbicides had arrived by the mid-1900s but electricity. Before electricity reached farms, chores aplenty were also still being done by hand. If something had to be done and it could only be done by hand, doing it that way stayed profitable because no one else could do it much faster. *Many* hand tasks played a role in keeping family farming viable. But clearly it was harvesting corn by hand that played the largest role.

Chapter Eleven

From Pioneering to Persevering

> THE RELENTLESS DYNAMIC of market growth [consolidated] disadvantages as well as advantages. To lose one's access to property at a time of rising land prices is to alter fundamentally one's long-term prospects. Historians have named this process one of Europeanization because the division between rich and poor, which had been modified during the initial period of settlement, reappeared.
>
> —Joyce Appleby, *Capitalism and a New Social Order*, pp. 10–11

The historian Joyce Appleby wrote the above passage about the thirteen colonies. She was describing the environs of busy port cities on the East Coast during the late colonial era of the mid-1700s, a time when those cities' environs were rapidly commercializing. Later, when rapid commercialization reached Indiana in the mid-1800s, its effect was the same: it undermined the egalitarian ethos of the pioneer era and divided Hoosiers more sharply on class lines.

As we saw in Chapters Six and Seven, central and western Indiana became quite heavily settled in the 1830s. Then both those parts of the state quickly commercialized. As that happened, letters written by some of their settlers to folks back home revealed new kinds of problems that were manifesting as settlement grew thicker.

Earlier letters from pioneers who wrote home were often exuberant, such as an 1810 letter from a Kentuckian who had just moved to Indiana's Knox County that year. "In this Neighborhood," Hosea Smith wrote his family back in Kentucky, "there is land aplenty not Entered and the richest I ever beheld and as for stock it is one of the finest places that is to be found." Hosea Smith bought 200 acres, not from the government but instead "with a Considerable Improvement with a crop in the ground." Land thus improved cost him only $3 an acre, merely a dollar more than the price of unsettled government land. Smith wanted all his Kentucky relatives to come. He wrote, "I am satisfied they might better themselves if they had not one Dollar after getting hear."[1]

Similarly, a newcomer named Caleb Lownes wrote home in 1815 that "the Soil and Climate so far as I have seen is in a Word delightful." He reported "I have taken a farm . . . about one Mile from town [Vincennes]—200 acres of which is a complete green grass pasture for sheep—150 acres of it is ploughing ground which last year yielded 75 bushels to the Acre [Lownes was sowing spring wheat, like the French]— favorable seasons it yields 80 bushels unfavorable 50— Oats from 30 to 40 bushels— On the edge of the prairie is the farm house, & 50 acres of orchard . . ."[2]

Close by, in Daviess County near the forks of White River, an English traveler wrote in 1819 that he "was passing through land as rich as a dung hill."[3] And as noted earlier, one hundred miles up the Wabash River in Parke County, a child in a family which arrived three years later (in 1822) recalled, "After the first year I never saw any scarcity of provisions."[4] Similarly, four years after that, in 1826 in the next county to the north (Fountain County), the family of John Wilkinson found the land so fertile at first that they harvested 60 to 100 bushels of corn an acre. They sold their considerable surplus for six cents a bushel to New Orleans-bound boatmen. The Wilkinsons had leased that farm for nine years but they paid off their lease after only four years; then they sold their other five years' right to use the land for $400 and moved further west, into Illinois.[5]

By 1830, however, a sterner economic challenge was starting to face many latecomers in areas that were already well-settled, especially if those latecomers arrived without a nest-egg. That year the Parke County latecomer William McCutcheon, lacking money, hired out his labor for wages, some of which he received in-kind. He wrote his father, "I hav had to work out for all the wheat we got since we come hear and for a cow and calf and other things or I wood been out of money before now wich keep me back in giting my farm open."

McCutcheon did manage to raise nine acres of corn for himself and half an acre of flax. He wrote his father, "if we have luck we will soon have a nice start." But when his hogs got into a clearing where he was burning logs and brush, eight of them did not get back out alive, and four others managed only barely. So rather than have hogs to sell in the fall of 1830, he would barely have enough for winter provisions.[6]

As the years passed and the presence of settlers thickened, letters written by settlers to their families and friends back home often revealed, like this last letter, a trend away from easy-entry farming. Most settlers did continue to achieve economic gains, but their letters often revealed that their problems were multiplying. In the 1820s and especially the 1830s, places that were becoming thickly settled saw the daily activities of well-off pioneers start to veer away from the daily activities of poor pioneers. Earlier, almost all pioneers had lived at roughly the same level, a level close to the ground. Although some early pioneers had possessed far more assets than other pioneers, almost all of them had faced heavy workloads. As the historian Jack Temple Kirby says, "For at least the equivalent of a generation, families performed virtually all the labor of production, and rural folk not only occupied much the same social class, but cemented communities through family labor exchanges. First generations of middle westerners, then, might be said to have lived *in* a market world, but were not *of* it."[7]

But then in the 1820s and particularly the 1830s, differing degrees of wealth began to show more visibly in people's daily lives. Pioneers with ample money could buy the land they wanted to live on. They would not have to watch their hard-earned improvements fall into someone else's ownership at a public land auction. Poor people, however, often *did* watch this happen, and the land's new owner might turn out to be a "low-down speculator" who refused to pay them for their improvements (buildings, land-clearing, fences, etc.). And even if they did receive payment for their improvements, the luckless poor often then faced a choice between becoming tenant farmers or else moving further west and starting over again on new land.[8]

Earlier, pioneers, both rich and poor, had often shared a sense of exuberance that was inspired both by the land's fecundity and by their joy in personally owning it. What a Wisconsin pioneer put on paper in 1842 had already been felt by thousands of Hoosier settlers: "Each night found us utterly fatigued. But we were exultant—it was all ours, and we were converting sheer wilderness into a smiling habitable land. There is a deep thrill of such effort—it causes a leaping of the heart incomprehensible to one who had not had the experience."[9] Such exuberance anticipated a future rooted in a family homeplace that would grow pervaded by fond memories. But that was a future which poor people were increasingly unable to reach. If they could not pay the price of their preemption stakehold, their "claim," then the fruits of their work would be reaped by others who *could* afford the price, whether those others were better-off settlers or were speculators. In fact, those two categories of better-off people were often one in the same, and they found their poorer neighbors increasingly willing to work for hire. Poorer men and boys were increasingly available, in exchange for wages, to do the most strenuous work—work like clearing land, plowing prairie, splitting rails, building fences, planting, cultivating, harvesting, building houses and barns, digging wells and lining them with stones, "even develop[ing] a whole farm complete with livestock ready to go into production."[10]

In 1854, for instance, the well-to-do ex-Congressman Elisha Embree wanted to have a track of woods that he owned in southwestern Indiana turned into a farm. So he leased it free for seven years to a man named James Minnes and stipulated that besides clearing and cultivating fifteen acres, Minnes had to use timber from the land to build a first-rate house with a brick chimney, and also had to build a stable, a smokehouse, and a corn crib, not to mention digging a well and furnishing Embree with split rails from the rest of the timber cut off the fifteen acres. Rather than being paid anything for his seven years of work, the tenant James Minnes merely gained the right to live off the farm that he was creating.[11] Through such exertions, many thousands of recent immigrants to America as well as native-born poor people tried to inch their way onto the "agricultural ladder" of upward mobility. Over 20 percent of Indiana's residents whose occupation in 1860 was "agricultural" were farm hands.[12]

Likewise, poorer women and girls were available to help with all the indoor work on farms, and also with milking the cows. Women's work was particularly heavy during spring and summer, especially if they had to cook and launder for hired hands as well as for their employer's family. In addition, hired girls often looked after the younger children of their employers.

As for poorer children, frequently they too were sent to work, hired out to help other families do many tasks, among them the planting of corn for which in the mid-1840s they were receiving twelve and a half cents a day. Besides the monotony of that, other drudgery awaited poor boys in their teens when they were old enough to rake out pigsties, load and spread manure, and split rails. The workdays often began at 4 A.M., even in wintertime, and lasted past dark.[13]

In one case in west-central Indiana's Clay County in the late 1840s, a boy performed heavy chores both morning and evening in exchange merely for room and board and miscellaneous expenses. His motivation was to live within walking distance of a school. Since the school was nonetheless five miles from the farm where he boarded, he had to get up very early to do his morning chores before walking to school.[14]

Back in the Introduction, when I first mentioned easy-entry farming, I said that the 1830s, '40s, and '50s were the decades when easy-entry farming ended in Indiana, ending first in fertile areas that lay close to navigable waterways or to cities, and then ending in the 1850s across most of the rest of the state because, by then, few nooks or crannies of Indiana still remained more than twenty miles away from a railroad or canal.[15]

Easy-entry farming had required three preconditions: (1) the needed skills had to be taught free of charge to rural children, (2) the needed land had to be cheap, and (3) the needed tools had to be cheap as well. In the 1850s the third of these preconditions became threatened by the rising cost of up-to-date farm implements, especially in wheat-growing areas[16]—such as east-central Indiana was fast becoming after midcentury. For instance, at the African-American "Beech" settlement in Rush County, the value of farm machinery per farm more than tripled between 1850 and 1870.[17] Overall, the total value of farm equipment on Indiana's farms rose 56 percent during the 1850s, although when calculated per farm it rose only from an average of $71.27 worth per farm in 1850 to an average of $87.22 worth in 1860. But keeping pace with the productivity gains on other farms did require that farmers spend more money to buy machinery. And by 1870 the value of farm implements would escalate to an average per Indiana farm of $109.60.[18]

Precondition number two (that the needed *land* be cheap) was threatened even more dramatically in the 1850s. That was when Indiana's land prices started to soar. In east-central Indiana, as throughout the state, many 40-acre homesteads had initially been bought from the U.S. government in the early to mid-1830s for $50 total.[19] And still as of 1850 the average value of farmland in east-central Indiana was only about $15 an acre. But by 1870 the average value of farmland there would rise to $50 an acre.

Granted, the rise in farmland values was not equally sharp everywhere around Indiana, but the state's overall average did go up from $10.66 an acre in 1850 to $21.73 in 1860 (and then to $35.03 an acre as of 1870).[20] Almost all of that land had cost next to nothing when it was first bought from the federal government. What had happened was explained in a nutshell by an observer in 1868: "By reason of the increased facilities furnished by railroads, for the transportation of agricultural products and cattle to the Atlantic cities, they were worth from 50 to 200 percent more in the Western States in 1860, than they were in 1840; and lands were raised in value in a corresponding manner."[21]

In fact, the rise of Indiana land prices in the 1850s and '60s actually outstripped the rise of corn, wheat, and pork prices. A pattern emerged of the smallest landowners selling out, the middling landowners neither gaining nor losing acreage, and the larger landowners buying up additional land.[22] As one owner of considerable scale (393 acres) in the Whitewater Valley wrote to a friend back home in 1856, "In a pecuniary point of view, I have done much better than I could have done in that country [Rockbridge County, along Virginia's Blue Ridge front]. I now own 393 acres of land which I could sell at 50 dollars per acre."[23]

Of the three preconditions that I listed for easy-entry farming, the first was the passing along of farm know-how from elders to youngsters—and that precondition would continue to stay firmly embedded for another 100 years. But let's glance at why the second and third preconditions for easy-entry farming had already ended by the 1850s or were rapidly doing so. The second precondition was cheap land and the third was cheap tools. The short answer to why they were ending is because suddenly not just a small *minority* of Indiana farmland could be commercially cropped to reap profits from the high prices that staple foods were fetching in far-off places where the new railroads and even newer trans-Atlantic steamships could carry them cheaply. The 1850s saw so many railroads built that suddenly not just farmland near cities or near navigable waterways could be commercially cropped with profit. Now *most* Indiana farmland could be commercially cropped for significant profits.[24] As the railroad-track mileage in Indiana leapt from a mere 226 miles in 1850 to 2,163 miles in 1860, the state's amount of "improved" farmland leapt by an astounding three million acres, from 5,046,543 acres in 1850 to 8,242,183 acres in 1860.[25] (The 1850s leap in Illinois's improved farmland was even more striking, starting out in 1850 below the Indiana acreage and by 1860 exceeding 13 million acres.)[26]

And note too that the investors who controlled the railroads back then before the Civil War were not yet the professional managers who controlled them after the Civil War and refined the gouging of farmers into a fine art.[27] As of 1860, farmers and drovers who sold live hogs in Cincinnati were still receiving 80 percent of the eventual value of the barreled pork and lard; and meanwhile at Chicago in 1860, farmers were still receiving 84 percent of the wholesale value of the Chicago market's nine leading farm commodities.[28]

A window of opportunity had opened in the 1850s because transportation breakthroughs gave Hoosier farmers new market access for their output, and that access was not yet controlled by monopolists. Most farmers were not yet constrained to sell their products cheaply to a railroad monopolist in time to make their mortgage payment to a money monopolist. On the money front, in fact, the 1850s were the decade of easy-entry banking, the "free banking" years when bankers had to compete avidly with each other to get business.[29] It is little wonder, then, that farm families who were doing better outnumbered farm families who were doing worse. In livestock, the average farm gained one additional horse and one additional head of cattle during the 1850s. This was only natural, since the average size of a farm rose from 137.8 acres to 150.7 acres, and the average number of *improved* acres per farm rose from 53.7 acres in 1850 to 65 acres in 1860.

Then, however, a reverse trend struck in the 1860s and struck so sharply that by 1870 the average farm size dropped way down to 118.8 acres. The improved acreage fell a bit too, down to an average of 62.8 improved acres in 1870. And meanwhile, as historian Clifton J. Phillips explains, "Indiana's farmers . . . found that their own rising production together with the recovery of European agriculture soon began to depress farm prices."[30]

So the 1850s had opened a window of opportunity for many farm families but that windowwas soon shutting again. *Why* were the 1850s that window, that fleeting golden age of family farming? Why not an earlier decade or perhaps a later one? The briefest answer is, because the prices of farm products were rising in the 1850s, and quite fast.[31]

The reasons for those 1850s price rises—three out of the four main reasons—were transportation reasons.[32] First, the prices for farm products were rising because the downriver system of exporting Indiana produce was still viable in the 1850s: the river route southward was still in competition with the canals and the new railroads. Despite all the new railroads built in the 1850s, the number of steamboats plying the Mississippi and Ohio rivers continued growing almost as fast as in earlier decades.[33] Twenty-nine percent of the Midwest's exports and imports were still traveling via the river route to and from New Orleans as of 1853.[34] Thus railroad shippers still had to compete with downriver flatboat captains to handle the farmers' exports, and still had to compete with upriver steamboat captains to handle the farmers' imports.

Secondly, *also* still in competition with the new railroads for Indiana farmers' exports and imports were canal shippers and Great Lakes shippers. A full one-third of the Midwest's total west-to-east tonnage and east-to-west tonnage still traveled by east-west waterways as late as 1862.[35] Of course, the shipping of goods by water was seasonal. During the winters, when canals and rivers were closed by ice, railroads frequently raised their shipping rates.[36]

And third is the reason mentioned above, that the railroads were not yet consolidated into monopolies and managed by professionals. Instead, they were still ardently competing against each other for the farmers' business. Not until the 1870s would Indiana's railroads fall under the control of outside monopolists—principally under the Pennsylvania system, the New York Central system, and the B&O system.[37] After that, those and a few other big railroad systems would rule the roost and their competitors for farmers' business would be virtually eliminated, namely their canal and riverboat competitors. But *before* the Civil War, the competition among shippers was so acute that the charges per "ton-mile" (to ship one ton of freight one mile) fell dramatically. Back in the 1800 to 1819 era, overland teamsters had charged 30 to 70 cents a ton-mile, whereas in the 1850s steamboats and canals charged just one to five cents per ton-mile, and railroads charged just two to five cents per ton-mile. But later, after 1880 at the latest, no more significant drops in freight rates occurred.[38]

The reversal of fortunes which struck Indiana farmers following the Civil War was sharp. But its preconditions had started taking shape much earlier, and in fact the yeoman-loving Jeffersonians had inadvertently helped to advance those preconditions of adversity. Thomas Jefferson and his fellow Democratic-Republicans of the 1790s had set out to safeguard the country's new republican institutions. To do that they had championed both political and economic equality among all adult white males. They opposed all privileges based on birth or wealth among adult white males. That was why Jefferson in the 1780s had convinced his home state of Virginia to abolish primogeniture and to replace it with partible inheritance.[39]

As the logical complement of political equality, the Jeffersonians tried to foster economic equality by adopting laissez-faire economic policies. And under the conditions of the 1790s, laissez-faire policies did diffuse economic opportunity. But the historian Joyce Appleby shows that that was due to the 1790s rising grain prices and rapidly rising American grain exports to Europe, which the Jeffersonians hoped would continue indefinitely and would thereby keep postponing American industrialization. Appleby does not think Jefferson and his friends would have championed laissez-faire economic policies if they had foreseen America's industrial future and its class consequences. She thinks they embraced "the liberal position on private property and economic freedom" only because they were "unable to envision a day when the free exercise of men's wealth-creating talents would produce its own class-divided society."[40] Except for one cotton-spinning mill that opened in 1791 in Rhode Island, American industrialization did not amount to much until 1811, when a large cotton mill was built at Waltham, Massachusetts. Actually U.S. industrialization did not gain much momentum until the late 1820s. When it did, however, it helped split the Jeffersonian electorate in two, separating those who still remained economic egalitarians from those who did not. The latter were led by people getting richer or trying to; they split off from the Democratic-Republicans and called themselves first National Republicans and then Whigs, creating a party that militantly favored laissez-faire industrialization despite the economic inequality that by then it was clearly generating.[41]

But here let's be careful. Perhaps *gross* economic inequality was not necessarily a consequence of laissez-faire industrialization. Gross economic inequality, after all, did not darken U.S. skies much until after the Civil War, when the U.S. financial system was twisted by money interests to perpetually maximize their own further enrichment by exploiting both farmers and wage workers.[42]

Meanwhile, back in the 1850s and back on the farm, many Indiana farm families had enough acreage and grew enough crops to make at least some of the new horse-powered equipment pay off—to make the new plows and harrows, the new wheat planters, reapers, and threshers, the new hay mowers and rakes, the new corn planters and cultivators, pay off. If their land wasn't heavily mortgaged, farm families often did very well and their 1850s success continued well into the 1860s with the help of the Civil War.[43]

They did well because most of the profits from the 1850s enormous reductions in transportation costs accrued at first to farmers in the form of higher prices paid to them for their products. Back in 1839, before the transportation revolution had yet shown many results in Indiana, a bushel of Indiana corn had brought its grower only about 16 cents. In 1842, if corn was hauled to Chicago it brought 21 cents a bushel. By 1852 the on-site Indiana sale price was up to 32 cents—but it was not up more than that yet because most of Indiana's railroads were still just getting incorporated. By 1857, however, and despite the economic recession which existed because of that year's financial crash, farmers were receiving 45 cents a bushel for corn. Then came the Civil War's price bulge, up to 79 cents for a bushel of corn in 1864.

Following the Civil War, however, came the sharp reversal. By 1872 a bushel of corn brought only 35 cents. It stayed low for a whole generation, falling in the 1890s below 30 cents. In January of 1900, a bushel of Indiana corn was selling for just 28 cents. (From then on, however, it rose steadily higher and higher until 1920.)[44]

Why, after the Civil War, did the price of corn race toward the bottom and stay there for a whole generation? Partly because of what I mentioned—not so much that farm families were swimming in a laissez-faire river but that after the Civil War they were swimming against its current. By then America's monetary current was flowing toward the nation's metropolises. Farmers' reactions to their adversity manifested in the Grange, the Greenback Party, and the populist movement. And one major endeavor of those farmers' groups was to keep waterborne transportation in business so it could continue to compete against the railroads.[45]

But other causes also underlay farm families' woes in the late 1800s. One of those was simply the continuing *lure* of easy-entry farming, which kept pulling more people onto the land farther west no matter how many of them went bust and had to leave again. The 1890 census announced that the U.S. no longer had a frontier line, but nonetheless more U.S. homestead claims were filed after 1890 than had been filed before that date.[46] In Indiana itself, from 1875 all the way to 1940, the number of farms that existed at any one time stayed fairly steady at around 200,000 farms,[47] but that number stayed steady only because an enormous number of unmarried youth and whole farm families went further west, and another enormous number moved off Hoosier farms to towns and cities, or anyway entered industrial jobs.

Those who went further west often pioneered farms on the Great Plains, contributing their share toward overproduction. Already by 1860, Indiana was losing more out-migrants than it was gaining in-migrants.[48] The 1880 census revealed that 436,551 Hoosier-born people were living by then in other states, and the states they mostly lived in were Illinois, Kansas, Missouri, and Iowa in that order.[49] The supply side of the U.S. supply-demand food equation was already glutted by that year (1880), but a few years later the Dakotas as well would be opened to grain farming. One prominent editor later dated American agriculture's structural unprofitability to that era when large grain surpluses started to manifest—and yet the vast Dakotas would soon be opened to homesteaders and start marketing mountains of grain as well.[50]

After that, Hoosiers who tried to compete in the wheat market were at a disadvantage, and by 1899 Indiana's wheat output would fall to seventh rank among the states.[51]

Thus, although the number of Indiana farm families stayed fairly steady after the Civil War, that was partly because thousands of Indiana's rural sons and daughters were moving farther west, where typically they kept farming. So much food came east on railroads that the prices paid for staple farm products fell drastically for *all* farmers regardless of where they were. The western railroads did buy trans-Atlantic steamship lines and use them to ship surplus food to Europe, but the main reason they bought the steamship lines was to bring more immigrants *from* Europe—immigrants who would grow more grain on the Great Plains and thereby bolster the railroads' freight incomes.[52]

So those were three of the reasons for Hoosiers' farm woes in the late 1800s. First, financial control by big-money interests; second, railroad consolidation and the atrophy of river and canal competition; and third, national overproduction of food. But there was also a fourth major reason for those rural woes, a reason that wasn't as complimentary to farmers as were those first three reasons. (Complimentary? Yes, since what led to overproduction was farm families' commendable work ethic, and since their struggles against financial and transportation monopolies were likewise commendable.)

But that fourth reason for rural woes was destructive exploitation of the land itself, and here the culprits were farmers themselves. They could claim extenuating circumstances, but one of those was not that no one had told them. Indeed, farmers themselves saw plainly the soil depletion and erosion they were causing. They called it "skimming" or "mining" the soil. "Farmers came from exhausted farms in the East to exhaust their farms in Ohio, then expected to go on to Indiana and Illinois. When warned of their neglect of manure, they would reply, 'When my land needs manure I will sell it and remove to the new country.'"[53]

Nor did people question farmers' right to ruin soil if they owned it. Many people did join in the hand-wringing, however, and also in efforts for improvement. Soil abuse helped prompt Indiana's General Assembly to start a state board of agriculture in 1851. By the end of 1852 forty-five county agricultural societies were operating, almost all of them either brand-new or newly revived. Annual state fairs started in October of that same year (1852), as did twenty county fairs. In 1853 Governor Joseph A. Wright (who also headed the state agricultural board) told the General Assembly that "the large surplus of pork grown in Indiana demonstrates the fact that our farmers raise too much Indian corn in proportion to the number of acres of improved land. This system of farming, if not modified by the introduction of diversified agricultural pursuits, must result in injury to the soil and loss to the farmers."[54]

It was already doing both. As Governor Wright also said, "If we keep on for half a century in the same mode of cultivation that some of us have pursued for the last few years, in some portions of the State, we shall not be able to raise a mullen stalk."[55]

But with prices for corn and pork going up in the 1850s and 1860s, quick profits took precedence over sound husbandry in most farmers' choices. Then later, during the overproduction glut of the later 1800s, when prices for farm products were falling, most farmers felt trapped in their farming system. By then it had become persevering that took precedence over sound husbandry.[56]

Meanwhile, mining the soil kept taking its toll on Indiana's farmland. Far from heeding Governor Joseph A. Wright's advice by planting fewer acres in corn, Hoosier farmers by 1878 had 4,215,000 million acres planted in corn, compared to only 2,864,556 million acres in 1870. Average corn output per acre meanwhile fell from 39.5 bushels in 1870 to only 32.8 bushels in 1878. And the price of corn fell from 64 cents a bushel in 1870 to only half that much in 1878, just 32 cents a bushel.[57] Many farm families' earlier race to maximize their land ownership and their future prospects had now been replaced by a gritty struggle simply to persevere.

But yes, if we look at only the 1850s, the picture looked quite hopeful for most Hoosier farm families. Their fortunes were still climbing. But meanwhile complications and challenges were multiplying, and a dark cloud was approaching from the west which would soon pour forth a torrent of overproduction. That is why, after the Civil War, farming for many Hoosiers would become a chronic struggle simply to persevere on the land. Not until the dawn of the new century in 1900 would the tide again turn in favor of farmers.

Epilogue

What Was at Stake?

I OFTEN THINK today of what an impact could be made if children believed they were *contributing* to a family's essential survival and happiness. In the transformation from a rural to an urban society, children are—though they might not agree—robbed of the opportunity to do genuinely responsible work.
—Dwight D. Eisenhower, *At Ease*, p. 33

Only those who love farming continue; the life requires commitment and is hazardous, demanding, and risky.
—Sonya Salamon, *Prairie Patrimony*, p. 38

What was at stake? What was gained, and lost, when family farming became virtually a relic of Indiana's bygone days?

Gained was more money—although not necessarily more money in the long run for most farmers. And what has been virtually lost is an early American way of living close to the earth in collaboration with neighbors. Rural Hoosier tradition was rooted in the self-sufficient and neighborly way that families once supported themselves and each other from the land. That tradition was not *exclusively* Hoosier, but in the Midwest it has been Indianans who have kept it alive the most, especially southern Indianans. When the Great Depression struck in 1929 and lasted throughout the 1930s, those Hoosier traditions proved very handy, and they will prove handy again if people lose their cash incomes again.

After a lifetime of pondering midwestern changes as they unfolded in the larger context of America and the world, the historian William N. Parker concluded that "the child born and growing up in the small family of locality and neighborhood acquires a certain character structure which the child growing up in the large institutional family of urban mass and bureaucratic organization does not acquire. That is why the distinction between rural and urban society and character has proved so fundamental in history. The improvement of means of communication has smudged the distinction, but the distinction remains between an environment in which relations are close and immediate and one in which they are extensive and remote."[1]

Two hundred years ago, Indiana was like a fountain of the Upper South's folk culture surging northward. The lifeways and values of upland Virginia, North Carolina, Tennessee, and Kentucky were being brought north and were saturating southern and central Indiana. The forebears of those "butternut" culture bearers had earlier come from Europe to British America, many of them as indentured servants. They had trekked southwest through Virginia's Shenandoah Valley and beyond. North Carolina's Piedmont, East Tennessee's Great Valley, and Kentucky's Bluegrass had each held them a while, but north across the Ohio the grass looked greener.

Their frontier farms demanded continuous labor and they parented numerous children. Thereby, their land hunger too was reproduced. Many had followed Daniel Boone's lead out of the Carolina Piedmont and through the Cumberland Gap into Bluegrass Kentucky, a base of operations from which they contested against Native Americans to control the rich soil of Ohio and Indiana. From the 1770s until the 1794 Battle of Fallen Timbers, the rivals raided each other back and forth across the Ohio River until, after 1794, the southerners flowed northward. The Michigan Road, which later bisected Indiana from the southeast to northwest basically extended the Wilderness Road that Daniel Boone had blazed through the Cumberland Gap into Kentucky.[2]

Today Indiana remains the most southern of midwestern states. Indeed, its incoming "butternuts" were already self-consciously southern two hundred years ago. In the 1820s they were telling each other that Ohio had been appropriated by Yankees but Indiana could be *their* place.[3] Now we know the outcome was a compromise, for southern Indiana is still culturally southern, but the state shades culturally northern as its latitude climbs. And by now southern versus northern seems less pressing than trying to understand what was at stake later when family farming started to falter.

Family farming was basically the same in both North and South. Rural self-sufficiently has been called culturally southern but actually it was barely more southern than northern. And market-oriented farming has been called culturally northern but it really was barely more northern than southern—as we saw in Chapter Five when tracing back the corn belt's origins.

In the big picture, what happened before 1900 included at least two revampings of the entire context that surrounded Indiana agriculture. Back when Native Americans had set the context, they lived in small, intimate membership groups ("bands") and their environment was a series of discreet commons, each under the jurisdiction of a specific band of people with specific membership.

Jurisdictions potentially stronger than Indian bands reached the future Indiana when European empires' agents and subjects arrived. Later came agents and citizens of the new American republic and the pace of jurisdiction change quickened. United States agents brought jurisdiction supposedly stronger than Indian jurisdiction but U.S. agents often exercised their authority half-heartedly so as to retain the support of U.S. citizens, especially those who resided nearby. Under U.S. jurisdiction, U.S. citizens wielded more rights than Indians—not just more civil rights but more economic rights.

America's national commitment to its citizens' individual economic opportunity grew during the early 1800s, culminating while the national Pre-emption Act was in effect, 1841 to 1891. Under its preemption rules, settlers legally held the first right to buy whatever parcel of public land they had settled on.

In 1891 preemption rights were ended, and by then some downsides of agricultural individualism were drawing adverse attention. During the 1870s and especially the 1880s, such huge surpluses of farm staples sought buyers that their prices plummeted. Support for economic individualism eroded among farmers and they banded together in common causes. Later, the 1930s depression eroded farmers' economic individualism yet further. In effect, the degree of micromanagement that Native Americans had once practiced was reinvented by the U.S. Department of Agriculture when the New Deal started in 1933.[4]

Fields within the study of history can be ranked by the size of the questions they try to grapple with. The past few decades have made ever clearer the reasons why traditional family farming is in a tailspin. It is easier and easier to see which traits of the U.S. economy and which policies of the U.S. government are undermining family farms' viability. Meanwhile, the reasons why family farming is such a dynamo of desirable human values have also grown more obvious with each survey of dysfunctional American families and each new report of schoolhouse mayhem. What is still unclear is only why so little is done to save family farms. Many American agricultural policymakers grew up on family farms and treasured that childhood experience. But the presumed economic trade-offs—the narrowly conceived "costs" of family farming—apparently loomed larger in their minds than childhood memories. Against small-scale farming, policymakers weighed not just production efficiency but also middle-class consumer amenities that they wanted farm families to be able to share with middle-class urbanities and suburbanites. Mary Neth thinks that by 1940 the die had already been cast. Before 1940, she says, "many voices debated the future of American agriculture, but the power of government and agricultural corporations and the hegemony of the social and economic ideal of progress overwhelmed alternative voices."[5]

Policymakers gave a green light to the slashing of per-unit production costs regardless of the social costs, and now we are wondering what kind of society we have left. Family farming had provided the main opportunity for boys to work side-by-side with their fathers, and nothing has replaced the crucial moral influence of father-son mentoring in the lives of American boys. In millions of households, girls do still work at least sometimes alongside their mothers, but relatively few boys can still do so with their fathers.[6]

What choice did policymakers have? Well, the competitive success of the Old Order Amish is no secret in Indiana or in the United States generally. Andrew Cayton and Peter Onuf point out in their book *The Midwest and the Nation* that "the triumph of commercial capitalism and the rise of a Midwestern bourgeoisie were not foreordained."[7] In this book I have tried to identify some choices that led toward today's rural Indiana, starting in Chapter Three with Thomas Jefferson's design of squared land parcels as a grid to superimpose on the Old Northwest, but also including most early settlers' decision to place their homesteads where economic advantages predominated, not where social advantages predominated.[8]

Either way, farming would have remained an easy-entry enterprise in at least some parts of Indiana until the mid-1800s, as it did. But if more pioneer families had placed their homesteads near other families, their daily lives would have been more social, and that might have fostered collective answers later when rural economic adversity struck. In turn, collective answers to rural economic adversity would plausibly have led to small-scale self-employed farming opportunities for wage workers when and if, as Jefferson put it in 1805, "it shall be attempted by the other classes to reduce them to the minimum of subsistence."[9] Jefferson hoped that easy entry into farming would continue in the U.S. to the thousandth generation.

The pioneer era's easy-entry farming depended on three preconditions, which I set forth in this book's Introduction and then glanced at again in Chapter Eleven. Those preconditions were (1) that the necessary skills be taught to rural children while they grew up, (2) that the necessary land be available cheaply, and (3) that the necessary tools be cheap as well.

Of those three preconditions, number one stayed true far longer than the other two. Even after World War II, many farm children were still learning the trade.[10] This was a source of American social strength. But the other two preconditions (easily available land and tools) started ending in Indiana in the mid-1800s. We have seen what happened to land prices in the 1850s when a profusion of railroads were built. (Land prices promptly soared.) As for cheap tools, we have seen the steady rise of farm-implement costs that started in the 1840s.

When the Civil War came, that gave financial interests in the East a chance to gain control over the banking laws of the nation as a whole and to twist them in their own favor.[11] In turn that then became one of the four main causes of the rural adversity that soured the late 1800s. The other three causes were transportation and storage monopolies, excessive farm output, and the degradation of soil. Indiana farmers tried to regulate and also to bypass monopolies, working through the Grange,[12] the Greenback Party, the populist movement, and countless other collective efforts, but nonetheless their defeat appeared total when William Jennings Bryan lost the 1896 presidential election.

Then, unexpectedly, the late 1890s brought worldwide gold discoveries, which dramatically increased the U.S. money supply. For the next twenty years, the status of farmers in the country's overall economy became so favorable that the best of the years that preceded World War I (1909–1914) were later enshrined as "parity."[13] So an upbeat mood returned to Hoosier farms in the early 1900s and bore convivial fruit in that era's Country Life movement and Good Roads movement, and in the willingness of farmers to even help pay the salaries of county agents assigned to teach them "book farming."

Besides the increased money supply that arrived on the eve of 1900, another reason why farm prosperity returned was because the growth rate of U.S. farm output had finally slowed down and the U.S. population's growth rate was finally outstripping it. So the twenty years from 1900 to 1920 saw the total income of U.S. farms more than double and the value of the average American farm more than triple—until the 1920s again brought dark clouds that lasted until World War II.[14]

But the 1900s lie beyond this book's scope. Suffice it to say that whereas in 1820 over three-fourths of Americans still lived on family farms, and even in 1900 over 39 percent still did so,[15] today less than 2 percent still do so and many of those are farming only part-time.

Does that make Indiana's family farm life of merely academic interest? No. And not just because some Indiana families still farm (by choice) but also because—who knows?—maybe Americans will someday heed the logic behind what France's minister of agriculture told the World Trade Organization in 1999: "Defending the family farm is the choice of a whole society."[16]

Notes

Introduction

1. See Parker, *Europe, America, and the Wider World*, vol. 2, pp. 249–50.
2. In the army from 1941 to 1945, Parker rose to the rank of major. *Yale Bulletin & Calendar*, 23 June 2000.
3. Schery, *Plants for Man*, pp. 368–69.
4. Parker, *Europe, America, and the Wider World*, vol. 2, pp. 130–36; and Carmony, *Indiana, 1816–1850*, pp. 66–68.
5. Parker, *Europe, America, and the Wider World*, vol. 2, pp. 235–36; and Walton, "River Transportation and the Old Northwest Territory," pp. 230–34. Overall in the five states of the Old Northwest by 1853, only 29 percent of exports and imports still traveled via New Orleans at the mouth of the Mississippi River. (Earle, "Regional Economic Development West of the Appalachians," p. 181.) Even from Cincinnati, by 1860 the amount of pork (including bacon) that was being shipped by canals and railroads had risen above the amount that was being shipped "south by water." Walsh, *The Rise of the Midwestern Pork Packing Industry*, p. 34 table. Also see John G. Clark, *The Grain Trade in the Old Northwest*, pp. 46–51, 212–36.
6. Atack and Bateman, "Yankee Farming and Settlement in the Old Northwest," p. 95.
7. Gates, *Landlords and Tenants on the Prairie Frontier*, pp. 13–47; and Gates, *The Farmer's Age*, pp. 11–12, 14–15. Specifically on Thomas and Nancy Hanks Lincoln's land woes in Kentucky, see Louis A. Warren, *Lincoln's Youth*, pp. 12–13. On Kentucky land titles generally, see Perkins, *Border Life*, pp. 119–135; and Abernethy, *Three Virginia Frontiers*, pp. 64–68, 76–82. On Tennessee title problems,

see Abernethy, *From Frontier to Plantation in Tennessee*, chapter 11; and Rankin, *Abolitionist*, p. 4. For earlier insecurity of titles in upland Virginia, see Hofstra, "Land Policy and Settlement in the Northern Shenandoah Valley," pp. 105, 111–12, 115–16.

8. U.S. Census of 1850, *The Seventh Census of the United States: 1850*, pp. xxxvi–xxxvii.
9. Atack and Bateman, *To Their Own Soil*, p. 74 table. Other nativities of Indiana household heads in 1860 included 13 percent born in Ohio and 8 percent born in Pennsylvania, but only about 8 percent born in any other northern state. Nineteen percent had been born in a foreign country.
10. Gates, *Landlords and Tenants on the Prairie Frontier*, pp. 144–48; and Stoll, *Larding the Lean Earth*, p. 30.
11. "The Hay Mill, A New Invention," in the Indiana Federal Writers' Project/Program Papers, Dearborn County section, 170 Agriculture, Indiana State University Library, Special Collections. (How half of Indiana's counties each became the subject of a county-history manuscript of up to 100 pages, written for the New Deal's Federal Writers Project, is explained by Blakey, *Creating a Hoosier Self-Portrait*, chapter 6.) For more about southeastern Indiana's early hay farms see John Edward Young, 1843, in McCord (comp.), *Travel Accounts of Indiana*, p. 186. The workings of early home-made hay presses powered by horses is explained by William Henry Smith, *A History of the State of Indiana*, p. 353. Starting in the East in 1853, factory-made hay presses became available. For their power, they remained dependent on horses until the mid–1880s, when hay presses powered by steam engines appeared. Hurt, *American Farm Tools*, pp. 95–96 illustrates and describes those.
12. Ellsworth, *The Valley of the Upper Wabash*, pp. 62–68, where a cost-benefit analysis of that hay business is also set forth.
13. How loose hay was lifted off wagons into barn lofts by horse-pulled ropes was described in 1996 by Alois Best of Floyds Knobs, which is located among hills not far from New Albany, Indiana and Louisville, Kentucky. See "Hoosier History: Down on the Farm," pp. 2–3. (More description, with illustrations, is in Hurt, *American Farm Tools*, pp. 92–94.) Much of southeastern Indiana remained in hay in the twentieth century, as shown by the 1955 regional "Farming Types" map in Dillon and Lyon, *Indiana: Crossroads of America*, p. 79.
14. Sorghum-growing was rapidly popularized in Indiana starting in 1856. (Esarey, *A History of Indiana*, vol. 2, p. 832.) All along, of course, maple syrup and honey also accounted for considerable sweetening.
15. Berry, *Western Prices before 1861*, p. 367. How the early millers, blacksmiths, and also hide-tanners performed their work is explained by Buley, *The Old Northwest*, vol. 1, pp. 225–28.
16. From early on, what farmers produced over and above their own home needs they called their "surplus." (Parker, *Europe, America, and the Wider World*, vol.

2, p. 170.) Agricultural historians have devised ingenious formulas based upon "corn-bushel equivalents" to measure the surplus of early American farms. Details are in Atack and Bateman, "Self-Sufficiency and the Marketable Surplus in the Rural North, 1860," and Atack and Bateman, *To Their Own Soil*, pp. 214–24.

17. Danhof, *Change in Agriculture*, pp. 35–38; and Buley, *The Old Northwest*, vol. 1, pp. 478–81.
18. Faragher, *Sugar Creek*, pp. 133–34; and Danbom, *Born in the Country*, pp. 90–91. Examples of such give-and-take in Indiana appear in many early accounts, such as Schramm, *The Schramm Letters*, p. 50. The continuing economic importance of neighborly reciprocity until at least World War II is documented by Neth, *Preserving the Family Farm*, chapter 2.
19. Esarey, *A History of Indiana*, vol. 1, p. 421.
20. Danhof, *Change in Agriculture*, pp. 90–94.
21. Danhof, *Change in Agriculture*, pp. 75–78; and Shannon, *The Farmer's Last Frontier*, p. 360.
22. Among many descriptions, see Gates, *The Farmer's Age*, pp. 86–89.
23. Fuller, *Summer on the Lakes*, p. 61; and Kirby, "Rural Culture in the American Middle West," pp. 590–92.
24. Varied interpretations of emotional religious revivals and worship services under frontier conditions are summarized by Cayton and Onuf, *The Midwest and the Nation*, pp. 47–49.
25. See Danhof, *Change in Agriculture*, pp. 2, 18–19 including n. 29. As regards lifespans, only 0.3 percent of rural Midwesterners in 1860 were aged 75 or over. (Atack and Bateman, *To Their Own Soil*, pp. 38–40 including graph.) In 1860 in Indiana overall (rural and urban both) only 0.021 percent of the population was aged 80 or over, only 1.02 percent was aged 70 or over, only 3.36 percent was aged 60 or over, and only 8.145 percent was aged 50 or over. U.S. Census of 1860, *Population of the United States in 1860*, pp. 110–11.
26. Hughes, "The Great Land Ordinances," pp. 3–8; Terry L. Anderson, "The First Privatization Movement," pp. 61–63; and Parker, *Europe, America, and the Wider World*, vol. 2, pp. 113–14, 170–71.
27. Hurt, "Midwestern Distinctiveness," p. 165.
28. Elkhart *Review*, 7 May 1859, quoted in Carter, "Rural Indiana in Transition," p. 108.
29. This book will not dwell much on land speculation from the viewpoint of speculators, but chapter 6 will describe one successful example set forth in Schramm, *The Schramm Letters*, pp. 55–56, 82–83.
30. Appleby, *Capitalism and a New Social Order*, p. 11. Also see Gates, *Landlords and Tenants on the Prairie Frontier*, pp. 5–6, 108–39.
31. Lincoln, "Annual Message to Congress," 3 December 1861, in Lincoln, *Complete Works of Abraham Lincoln*, vol. 7, pp. 58–59.

32. Lincoln, "Annual Message to Congress," 3 December 1861, in Lincoln, *Complete Works of Abraham Lincoln*, vol. 7, pp. 57–58.
33. Jefferson to a Mr. Lithson, 4 January 1805, quoted in Appleby, *Capitalism and a New Social Order*, p. 99; also see pp. 100–5 there.
34. Thomas Jefferson, 1785, quoted in Berry, *The Unsettling of America*, p. 220.
35. Thomas Jefferson, 1785, quoted in ibid.
36. Among many first-hand accounts is Jager, *Eighty Acres*. As of 1945, the U.S. still held 5.9 million farms, home to 24.4 million farm residents (17.5 percent of the nation's population) and the average size of those farms was 195 acres. But by the year 2005 the U.S. held only about two million farms and their size averaged about 450 acres. As for Indiana, its 176,000 farms as of 1945, home to 19 percent of the state's population, fell to less than 60,000 farms by the year 2005. Meanwhile, the intervening sixty years had seen Indiana's average farm size climb from 114 acres to about 255 acres. U.S. Census Bureau, *Historical Statistics of the United States*, Part 1, pp. 457–61; and U.S. Census Bureau, *Statistical Abstract of the United States, 2004–2005*, pp. 526, 528.
37. *Indiana Farmer*, 22 June 1878, as quoted in Argersinger and Argersinger, "The Machine Breakers," pp. 407–8.
38. Ibid., pp. 399, 404–7.

Chapter One

1. Jones, *History of Agriculture in Ohio*, pp. 10–11.
2. Bruce D. Smith, *The Emergence of Agriculture*, pp. 185–96. Also see Asch and Asch, "Prehistoric Plant Cultivation in West-Central Illinois," pp. 153–58, 196–203; and Hurt, *Indian Agriculture in America*, p. 11.
3. Hurt, *Indian Agriculture in America*, p. 9; Prince, *Wetlands of the American Midwest*, pp. 85, 87; and Shaffer, *Native Americans before 1492*, p. 25.
4. This wording is Douglas Hurt's summary of reported Cherokee views. Hurt, *Indian Agriculture in America*, p. 33.
5. Among many accounts, see Christian, *Maps of Time*, pp. 189–90.
6. Cowan, "Understanding the Evolution of Plant Husbandry in Eastern North America," p. 243.
7. A 1791 report quoted in a 1793 account by Gilbert Imlay, republished in Lindley (ed.), *Indiana as Seen by Early Travelers*, p. 13. Also regarding rituals and festivals see Beckwith, *The Illinois and Indiana Indians*, pp. 169–70; Burnet, *Notes on the Early Settlement of the North-Western Territory*, pp. 120–21; and John Johnston, in Hill, *John Johnston and the Indians*, p. 192.
8. Again, quoted by Imlay, in Lindley (ed.), *Indiana as Seen by Early Travelers*, p. 13.
9. One among innumerable accounts is Clifton, *The Prairie People*, pp. 121–25.
10. Shepard, Interview, in *Listening to the Land*, pp. 249–51.
11. Calvin Martin, "The Metaphysics of Writing Indian-White History," pp. 29–32;

Clifton, *The Prairie People*, pp. 121–23; Cass, *Considerations on the Present State of the Indians*, p. 12; and Cowan, "Understanding the Evolution of Plant Husbandry in Eastern North America," p. 224.

12. Calvin Martin, *Keepers of the Game*, pp. 113–30. Pennsylvania Indians reportedly believed that a *single* "Keeper of the Game" controlled the availability of *all* game to hunters. See Witthoft, *The American Indian as Hunter*, pp. 5–9.
13. For example, see Somé, *Of Water and the Spirit*, pp. 185–89, 302–9.
14. Cowan, "Understanding the Evolution of Plant Husbandry in Eastern North America," pp. 226–31; Sahlins, "The Original Affluent Society," pp. 11–14, 28–32. For estimates of hunter-gatherers' leisure time, see Sahlins, pp. 14–27; and Allen Johnson, "In Search of the Affluent Society," pp. 52–55, 56 graph, 58 graph.
15. Faulkner, *The Late Prehistoric Occupation of Northwestern Indiana*, pp. 36–37; and Stout, "Report on the Kickapoo, Illinois, and Potawatomi Indians," p. 301.
16. Glenn and Rafert, "Native Americans," pp. 392–94. For an excellent *continent-wide* description with slightly different periodization, see Salisbury, "American Indians and American History," pp. 47–51.
17. Shaffer, *Native Americans before 1492*, pp. 19–22. It is now known that settled communities likewise predated the domestication of plants or animals in the Middle East and in what's now the southern Sahara Desert. In each case, those pre-agricultural settlers had access to river or wetland wild food sources. See Bruce D. Smith, *The Emergence of Agriculture*, pp. 196–97, 210–12.
18. Ford, "The Processes of Plant Food Production in Prehistoric North America," pp. 2–7; and Christian, *Maps of Time*, pp. 224–38.
19. Bruce D. Smith, *The Emergence of Agriculture*, pp. 192–96. See also Asch and Asch, "Prehistoric Plant Cultivation in West-Central Illinois," pp. 153–58; and King, "Early Cultivated Cucurbits in Eastern North America," pp. 73–74, 79–80, 95–97.
20. Bruce D. Smith, *The Emergence of Agriculture*, pp. 11–13 (including map and graph), 185–96; and Pringle, "The Slow Birth of Agriculture," pp. 1446–50. Also see Asch and Asch, "Prehistoric Plant Cultivation in West-Central Illinois," pp. 153–83, 202. Drawings of lambs-quarter, marsh elder, knotweed, maygrass, and little barley appear in Justice, *Learning from Prehistory*, p. 51. See also pp. 48–49.
21. Shaffer, *Native Americans before 1492*, pp. 24–25, 44–46; Lambert, *Traces of the Past*, pp. 216–17; Galinat, "The Evolution of Corn and Culture in North America," pp. 541–43; Galinat, "Domestication and Diffusion of Maize," pp. 245–48, 264–70 (including map), 274–78; Wagner, "Corn in Eastern Woodlands Late Prehistory," pp. 342–43; and Baker, "Indian Corn and Its Culture." Note that cross-pollination between gourdseed and flint corn had earlier occurred in the East, by 1812 at the latest, but that crosses which occurred later in the Midwest were the ones that came to cover the corn belt. (Rasmussen [ed.], *Readings in the History of American Agriculture*, pp. 60–61.) Both gourd-

seed and flint corn—their plants, tassels, ears, and kernels—are illustrated in Hudson, *Making the Corn Belt*, pp. 51–52.

22. Shaffer, *Native Americans before 1492*, p. 45. For evidence that corn became *culturally* important in the Midwest centuries before it became a food staple there (which occurred between A.D. 750 and 1050) see Hastorf and Johannessen, "Becoming Corn-Eaters in Prehistoric America," pp. 428, 431–37, 439–43.
23. Lambert, *Traces of the Past*, pp. 216–17. The proportion of Indians' diet that consisted of corn can be estimated from the collagen of their bones because maize was the only tropical grass that they ate. Lambert adds that "these increases in a cultivated food permitted large increases in population. Disease patterns that may be discerned from the bone, however, also increased, possibly as a result of insufficient dietary variability."
24. Reidhead, *A Linear Programming Model of Prehistoric Subsistence Optimization*, pp. 201–2; and Yarnell, "Early Plant Husbandry in Eastern North America," p. 272.
25. Indeed, there's now evidence of bean cultivation in the Andes Mountains 2,000 years earlier than any domestic beans have been dated in Mexico. Bruce D. Smith, *The Emergence of Agriculture*, pp. 160–63.
26. Hurt, *Indian Agriculture in America*, pp. 6–7; and Kaplan, "Archeology and Domestication in American Phaseolus (Beans)," pp. 519–20. Corn's natural protein content is about 10 percent. Ebeling, *The Fruited Plain*, p. 177.
27. Asch and Asch, "Prehistoric Plant Cultivation in West-Central Illinois," pp. 187–90, 195–96; and Yarnell, *Aborigional Relationships between Culture and Plant Life in the Upper Great Lakes Region*, pp. 65, 79. The Potawatami are Algonkian, closely related to the Ottawa and Ojibwa. Regarding the Potawatomi and wild rice, see Berthrong, *Indians of Northern Indiana and Southwestern Michigan*, pp. 59–60.
28. Numerous historical sources document women's predominance in farming, including Beckwith, *The Illinois and Indiana Indians*, p. 169; and Henry Harvey as quoted in Brelsford, *Indians of Montgomery County, Indiana*, p. 134. For non-Indiana evidence and sources see Jensen, "Native American Women and Agriculture."
29. Reidhead, *A Linear Programming Model of Prehistoric Subsistence Optimization*, pp. 201–3. A contrary view—that most Indians practiced alternate-year fallows—is set forth by Doolittle, *Cultivated Landscapes of Native North America*, pp. 174–90.
30. Boserup, *The Conditions of Agricultural Growth*, chapters 1, 3, 5, 7, and 8.
31. The hours for land clearing, as prorated per year, would have been fewer if Robert Leslie Jones is correct that Indians in Ohio (and therefore plausibly in Indiana as well) moved their fields only every "five or ten years, or perhaps twice in a generation." Jones, *History of Agriculture in Ohio*, pp. 10–11.
32. Reidhead, *A Linear Programming Model of Prehistoric Subsistence Optimization*, pp. 203–6.

33. For a fuller discussion, with explanatory maps, see the *Handbook of North American Indians*, vol. 15: *Northeast*, pp. 547–68.
34. Recorded epidemics among Indians in or near Indiana include smallpox epidemics in 1733, 1752, 1757, 1762–64, 1781–82, 1787–88, and 1801. A measles epidemic occurred in 1715 and a scarlet fever epidemic in 1793–94. Glenn and Rafert, "Native Americans," pp. 395–96; and the *Handbook of North American Indians*, Vol. 15: *Northeast*, pp. 352, 387.

Chapter Two

1. Barnhart and Riker, *Indiana to 1816: The Colonial Period*, p. 62.
2. *Handbook of North American Indians*, Vol. 15: *Northeast*, p. 597; and Hinderaker, *Elusive Empires*, pp. 14–17 (including map dated 1684), 47–51.
3. Clifton, *The Prairie People*, pp. 125–28.
4. Paré, "The St. Joseph Mission," pp. 24–25; Robertson, *Valley of the Upper Maumee River*, pp. 49–51; and Barnhart and Carmony, *Indiana: From Frontier to Industrial Commonwealth*, vol. 1, pp. 19–21.
5. Berthrong, *Indians of Northern Indiana and Southwestern Michigan*, p. 33; Benton, *The Wabash Trade Route*, pp. 16–19; James H. Kellar, *An Introduction to the Prehistory of Indiana*, p. 63; and Hinderaker, *Elusive Empires*, pp. 16–18, 47–50.
6. Reidhead, *A Linear Programming Model of Prehistoric Subsistence Optimization*, p. 203.
7. Allen Johnson, "In Search of the Affluent Society," p. 55.
8. See, for example, Benton, *The Wabash Trade Route*, pp. 17–26.
9. Robertson, *Valley of the Upper Maumee River*, p. 48. For 1731 as the date when Vincennes was founded, see Hinderaker, *Elusive Empires*, p. 50.
10. Barnhart and Carmony, *Indiana: From Frontier to Industrial Commonwealth*, vol. 1, p. 24. The Wea were a branch of the Miami people. Another Indiana-based branch of the Miami were the Piankashaws.
11. This is the 1791 report cited above in chapter 1, note 7. It is quoted in a 1793 account by Gilbert Imlay, as republished in Lindley (ed.), *Indiana as Seen by Early Travelers*, pp. 13–14. Corn "in the milk" is what is now called corn on the cob. Kenapacomagua was located six Wabash River miles above present-day Logansport, where it stretched for two or three miles along the Wabash-tributary Eel River. Cayton, *Frontier Indiana*, p. 158.
12. Quoted in Barce, *The Land of the Miamis*, p. 42.
13. Barce, *The Land of the Miamis*, pp. 42–43.
14. *Handbook of North American Indians*, Vol. 15: *Northeast*, pp. 30–31, 550; and Shaffer, *Native Americans before 1492*, pp. 58, 76.
15. Hodge (ed.), *Handbook of American Indians North of Mexico*, Part 1, p. 615.
16. Gilbert Imlay (1793), in Lindley (ed.), *Indiana as Seen by Early Travelers*, p. 11. On Kentucky and Ohio locations of early charcoal-fired iron-making furnaces, see the *Encyclopedia of Appalachia*, pp. 492–96.

17. Hurt, *Indian Agriculture in America*, p. xi.
18. McIsaac, "Sustainable Agriculture," p. 255.
19. Perkins, *Border Life*, p. 120. A map of the waterways within the Ohio River watershed appears in Perkins at p. 11. The number of Indians in the Old Northwest as of the late 1600s has been estimated at between 75,800 and 106,070. Prince, *Wetlands of the American Midwest*, pp. 88–91 including tables and map.
20. These four categories seem to more or less summarize what Europeans and Americans have usually wanted from indigenous people in all parts of the world for the past several hundred years. I've taken them from the "imperatives" that, according to Arturo Escobar, shaped U.S. policies toward the Third World following World War II. (Escobar, *Encountering Development*, p. 71.) It may be odd to draw on a study of today's Third World to summarize what Euro-Americans wanted from Indians in the 1600s and 1700s, but these four Western desires have arguably been affecting indigenous people's lives for several centuries now.
21. Wessel, "Agriculture, Indians, and American History," p. 13.
22. Glenn and Rafert, "Native Americans," pp. 394–95; and Cass, *Considerations on the Present State of the Indians*, pp. 5–6.
23. Among many accounts, see Calvin Martin, *Keepers of the Game*, pp. 151–54. Also Cayton, *Frontier Indiana*, pp. 262–63.
24. James A. Brown, "The Impact of the European Presence on Indian Culture," pp. 6–15, 17–19.
25. Cayton, *Frontier Indiana*, chapters 6–8.
26. Dennis Michael Warren, "Indigenous Agricultural Knowledge, Technology, and Social Change," p. 38.
27. Hinderaker, *Elusive Empires*, pp. 41–44.
28. Quoted in Hudson, *Making the Corn Belt*, p. 34. Also see pp. 31–34.
29. Hudson, *Making the Corn Belt*, p. 35 (also see pp. 36–39); and Thomas D. Clark, "The Advance of the Anglo-American Frontier," pp. 79–80, 86.
30. Hudson, *Making the Corn Belt*, pp. 36, 39–44. Other examples of using "old fields" in Ohio are in Jones, *History of Agriculture in Ohio*, pp. 11, 28. A map of the Virginia Military Tract and other land tracts set aside in Ohio appears in Robert M. Taylor, Jr. (ed.), *The Northwest Ordinance*, p. 77b.
31. Thompson, *Sons of the Wilderness*, p. 131; and Redmond (ed.), *Current Research in Indiana Archaeology and Prehistory: 1991 & 1992*, pp. 10–14.
32. Roosevelt, *The Winning of the West*, Vol. 1, pp. 69–71. A more nuanced interpretation of Indian land-tenure customs is found in Hurt, *Indian Agriculture in America*, pp. 65–69.
33. Cass, *Considerations on the Present State of the Indians*, pp. 16–48. Regarding tenacious European assumptions that underlay how Euro-American legal categories were applied to Native Americans, see Berkhofer, "Cultural Pluralism Versus Ethnocentrism in the New Indian History," pp. 37–39.
34. Banner, *How the Indians Lost Their Land*, chapter 5.

35. Glenn and Rafert, "Native Americans," p. 400. Later, the stubborn Mississinewa band were the very last of the Miami to move westward. Beckwith, *The Illinois and Indiana Indians*, pp. 113–14.
36. Quoted in Dillon, "The National Decline of the Miami Indians," p. 136. The official who penned this letter isn't named.
37. Dillon, "The National Decline of the Miami Indians," pp. 136–37.
38. Paul Weer, "Preliminary Notes of the Moravian Mission to the Delaware Indians on White River, Indiana," pp. 9–10, 12–13. In the Paul Weer Papers, Folder 10. Also see Cayton, *Frontier Indiana*, pp. 196–99, 205–9. Usually the number of Indian villages along White River is given as nine. For details see Thompson, *Sons of the Wilderness*, pp. 196–205.
39. Glenn and Rafert, "Native Americans," p. 397 map. Also see Esarey, *A History of Indiana*, vol. 1, p. 329 map.
40. Glenn and Rafert, "Native Americans," pp. 396–404 including maps; Edmunds, *The Potawatomi: Keepers of the Fire*, pp. 215–72; Buley, *The Old Northwest*, Vol. 1, pp. 18–19, 36–38, 99–105, 109–15 including map; and the Ewing Brothers Papers in the Manuscript Section of the Indiana Division, Indiana State Library, Indianapolis.
41. Quoted in Dillon, "The National Decline of the Miami Indians," pp. 136–37. Also see Thompson, *Sons of the Wilderness*, pp. 75–76, 81; and John Johnston, in Hill, *John Johnston and the Indians*, pp. 169–70.
42. Among countless such assertions, see Cass, *Considerations on the Present State of the Indians*, p. 17. Cass mistakenly assumed that "corn, beans, and pumpkins were indigenous to the country" (p. 5).
43. Regarding Moravians, see Gipson (ed.), *The Moravian Indian Mission on White River*, pp. 297, 450. Regarding Quakers, see Woehrmann, *At the Headwaters of the Maumee*, chapter 5. Starting back in 1789, Quakers who worked among Senecas in western New York State had begun a similar campaign to foster European-style farming, including there too the use of draft animals and the social dominance of men over women. See Jensen, "Native American Women and Agriculture," especially pp. 51–56.
44. Hurt, *Indian Agriculture in America*, p. 38.
45. Brelsford, *Indians of Montgomery County, Indiana*, p. 9. "Egypt" lay in Madison Township of Montgomery County. The Euro-American settlers' problem was that wild animals and birds consumed their corn in the fields. Among Indians, the job of preventing that was customarily assigned to children.

Chapter Three

1. Thomas Jefferson, 1787, quoted in Ekberg, *French Roots in the Illinois Country*, p. 263.
2. Sauer, "Homestead and Community on the Middle Border," p. 37.

3. Esarey, *A History of Indiana*, vol. 1, p. 421; Esarey, *The Indiana Home*, pp. 16–17, 24–25; and Buley, *The Old Northwest*, vol. 1, pp. 141–42. "Mast" as hog feed was such an asset to pioneer farmers that they sometimes climbed up into beech and oak trees in springtime to estimate their thickness of bloom, and then planted only enough corn to compensate for any shortage of beechnuts and acorns that they estimated would occur in the autumn. So claims Power, *Planting Corn Belt Culture*, p. 162. Corroboration appears in Nicholson, "Swine, Timber, and Tourism," pp. 29–30. A late-spring freeze could ruin the autumn's "acorn crop," says Nicholson.
4. Benton says that it was founded in 1727 and became known as Vincennes in 1752. Benton, *The Wabash Trade Route*, p. 19.
5. Wilson, "Vincennes," pp. 26–29. Moravian missionaries who visited Vincennes in 1792 estimated the size of the town plots at only about one-third of an acre each. (John Heckewelder, 1792, in McCord [compiler], *Travel Accounts of Indiana*, p. 34.) For a drawing of a typical such French house, see Buley, *The Old Northwest*, vol. 1, opposite page 49.
6. Benton, *The Wabash Trade Route*, p. 24. Two years later in 1748, the overall shipments of flour from Vincennes *and* the Illinois settlements were recorded as totaling 800,000 pounds (400 tons)—not to mention pork, dried beef, and other products. Briggs, "Le Pays des Illinois," p. 51.
7. Ekberg, *French Roots in the Illinois Country*, p. 155 table.
8. George Croghan, 1765, in McCord (compiler), *Travel Accounts of Indiana*, p. 19; Ekberg, *French Roots in the Illinois Country*, pp. 196–98, 205 (table), 271; and Benton, *The Wabash Trade Route*, p. 24. Benton adds that "under English occupation" the "Wabash posts languished" and their trade declined. Thus their production figures plausibly were higher before 1760. Benton, pp. 30–31.
9. Briggs, "Le Pays des Illinois," pp. 50–53; and Ekberg, *French Roots in the Illinois Country*, pp. 280–82.
10. Ekberg, *French Roots in the Illinois Country*, pp. 82–88. For the similar arrangements at Kaskaskia and Chartres on the Mississippi, see Briggs, "Le Pays des Illinois," pp. 40–47.
11. Ekberg, *French Roots in the Illinois Country*, p. 192. Details about how Vincennes residents coordinated the timing of their annual agricultural sequence are on pp. 119–23.
12. Ekberg, *French Roots in the Illinois Country*, pp. 178–85, 192–93. William C. Latta says that Vincennes' French settlers also "had good horses which they obtained from the Indians who brought them from the Spanish settlements west of the Mississippi." Latta, *Outline History of Indiana Agriculture*, p. 31. Also see Benton, *The Wabash Trade Route*, p. 24.
13. Esarey, *The Indiana Home*, p. 56; Benton, *The Wabash Trade Route*, pp. 22–23; and Thomas Hutchins, 1762, in McCord (compiler), *Travel Accounts of Indiana*, pp. 11–12.

14. Briggs, "Le Pays des Illinois," pp. 50–51; and Ekberg, *French Roots in the Illinois Country*, pp. 213–31.
15. Ekberg, *French Roots in the Illinois Country*, pp. 260–63. A contrary quantification of early Midwestern violence was set forth by Morris Birkbeck in his *Notes on a Journey to America* (1818). Birkbeck wrote of early-statehood southwestern Indiana that "drunkeness is rare, and quarrelling rare in proportion" (p. 108). Birkbeck did however assert that near the Ohio River "every hamlet is demoralized, and every plantation is liable to outrage" (p. 111). Logan Esarey says that as late as 1816 the Indiana side of the Ohio River was a by-word whose few residents were reputed to be "'hoss thieves' and other criminals who had run away from the older settlements." Esarey, *The Indiana Home*, p. 18.
16. Hofstra, "'The Extension of His Majesties Dominions,'" p. 1296. Colonial Virginia's application of such policies is detailed in Hofstra, "Land Policy and Settlement in the Northern Shenandoah Valley"; and in Hofstra, *The Planting of New Virginia*, chapter 2, especially pp. 68–93.
17. Quoted in Hofstra, "'The Extension of His Majesties Dominions,'" p. 1309.
18. Hofstra, "'The Extension of His Majesties Dominions,'" p. 1311. Also see Abernethy, *Three Virginia Frontiers*, pp. 39–57.
19. Hofstra, "Land Policy and Settlement in the Northern Shenandoah Valley," pp. 106–9, 112; and Abernethy, *Three Virginia Frontiers*, pp. 38, 60–61. Later, when the U.S. government distributed western land warrants to veterans of the Revolutionary War, that too was motivated (at least partly) by a desire to strengthen frontier security—as documented by Lebergott, "'O Pioneers,'" pp. 41–42. Many *landless* Kentuckians, in fact, declared themselves *un*willing to help defend Kentucky. Barnhart, *Valley of Democracy*, pp. 54–58.
20. Hinderaker, *Elusive Empires*, pp. 268–69. In 1779–1780, protests by common people in Kentucky and points west included a petition for separate statehood signed by 640 individuals. They were upset by Virginia's elitist Land Act of 1779. One Virginia leader lamented: "We have distressing news from *Kentucke* which is entirely oewing to a Set of Nabobs in Virginia, taking all the Lands there... Hundred of Families are ruin'd by it . . . It is a Truth that the People there publicly say it—Let the great men . . . who the Land belongs to, come & defend it, for we will not lift up a Gun in Defence of it." Quoted in Barnhart, *Valley of Democracy*, p. 54. Such protests led to liberal national land policies (pp. 54–58).
21. William Crawford to George Washington, 15 March 1772, in Butterworth (ed.), *The Washington-Crawford Letters*, p. 25.
22. Ekberg, *French Roots in the Illinois Country*, pp. 256–63. Another interesting interpretation of frontier violence appears in Cayton and Onuf, *The Midwest and the Nation*, pp. 70–73.
23. Parker, "From Northwest to Midwest," p. 14. Earlier, Frederick Jackson Turner bluntly asserted that "individualism in America has allowed a laxity in regard to governmental affairs which has rendered possible the spoils system and all

the manifest evils that follow from the lack of a highly developed civic spirit. In this connection may be noted also the influence of frontier conditions in permitting lax business honor . . ." Turner, *The Frontier in American History*, p. 32.

24. Quoted in Gitlin, "On the Boundaries of Empire," p. 86.
25. Freyfogel, *Bounded People, Boundless Lands*, p. 150.
26. Dennis Michael Warren, "Indigenous Agricultural Knowledge, Technology, and Social Change," p. 38.
27. "Taking Action on Hog Farms," *Yes!* Magazine, issue #30 (summer 2004), p. 5. More information is at the website of the Community Environmental Legal Defense Fund: www.celdf.org.
28. Esarey, *The Indiana Home*, pp. 67, 72.
29. Carter, "Rural Indiana in Transition," p. 107.
30. Lebergott. "'O Pioneers'," pp. 40–41.
31. "Colonus," pp. 482–83.
32. Lyon, *The Origins of the Middle Ages*, pp. 68–81.
33. Weber, *The Agrarian Sociology of Ancient Civilizations*, p. 323. On ancient Rome being capitalistic, pp. 315–335. On ancient Rome not industrializing, p. 393.
34. Cameron, *A Concise Economic History of the World*, pp. 51–54.
35. Lyon, *The Origins of the Middle Ages*, pp. 69, 71; and Claster, *The Medieval Experience*, p. 108.
36. Marks, *The Origins of the Modern World*, chapter 3.
37. Christian, *Maps of Time*, pp. 398–403, 413–18; and Marks, *The Origins of the Modern World*, p. 138. Not *just* the availability of low-wage workers fostered industrialization, of course, but also a spate of mechanical inventions. See Christian, pp. 418–26.
38. Among countless examples of this, a few Indiana ones are recounted in Schramm, *The Schramm Letters*, pp. 50, 64–65.

Chapter Four

1. More about Ohio River migrants of the 1780s is in Barnhart, *Valley of Democracy*, pp. 37–39. *One* supposedly legal settlement of Americans did exist north of the Ohio River in what would soon be designated Indiana Territory: "Across [the Ohio River] from Louisville, families had settled on 150,000 acres granted [in 1783] by Virginia to participants in the military expeditions of George Rogers Clark during the American Revolution." (Cayton, "The Northwest Ordinance from the Perspective of the Frontier," p. 5.) Six maps which chronicle the Old Northwest's six successive name-and-boundary changes between 1800 and 1818 appear in Buley, *The Old Northwest*, vol. 1, pp. 62–64.
2. Public land in "the Gore" slice of southeastern Indiana was sold at the land office in Cincinnati. "The Gore," had been ceded by Indians in 1795 under the Treaty of Greenville. Along the western edge of "the Gore," the "Twelve Mile

Purchase" of 1809 added another vertical strip of land which settlers could likewise purchase at the Cincinnati land office (In Shonkwiler, "The Land Office Business in Indiana," see the map of Indiana's land-office districts as of 1849 [following p. 10 there].) Buley, *The Old Northwest*, vol. 1, p. 117 (map) gives land-office openings as Vincennes in 1804, Jeffersonville in 1807, Brookville and Terre Haute both in 1819, Fort Wayne in 1822, Crawfordsville in 1823, Indianapolis in 1825, and LaPorte in 1833. But Rohrbough, *The Land Office Business* (pp. 28–29 map) gives 1810 as the date when the Jeffersonville land *district* was legally established. Maps showing the boundaries of the later public-land districts can be found on Rohrbough's pp. 128–29 and 236–37.

3. Birkbeck, *Notes on a Journey to America*, p. 79. Birkbeck fails to make clear that those land prices were close to the maximum. Typical Ohio land prices were lower. Jakle, *Images of the Ohio Valley*, pp. 98–99.
4. "Commerce, Industry, and Agriculture in Franklin County," p. 6, in the Indiana Federal Writers' Project/Program Papers, 170 Agriculture, Indiana State University Library, Special Collections. In fact, not just corn did better than wheat in rich new soil, but Indiana's other four small grains also did better than wheat in rich soil—namely, oats, rye, barley, and buckwheat. (Carmony, *Indiana, 1816–1850*, p. 53.) None of those four fetched high prices, but oats were commonly grown to use as horse feed.
5. Birkbeck, *Notes on a Journey to America*, pp. 90–91; and Rohrbough, *The Trans-Appalachian Frontier*, p. 166. The evolution of official U.S. policies for transferring public land to private ownership is chronicled in the *Encyclopedia of the United States in the Nineteenth Century*, vol. 1, pp. 458–60. More details appear in Rohrbough, *The Land Office Business.* A description of the surveying work appears in Buley, *The Old Northwest*, vol. 1, pp. 115–23 including map. In addition, detailed maps showing each Indiana county and its numbered sections as of 1876 appears in the *Illustrated Historical Atlas of the State of Indiana*.
6. Louis A. Warren, *Lincoln's Youth*, pp. 15–23. Today that land is in Spencer County. For a map showing the far southern counties of Indiana as of 1816, see Sieber and Munson, *Looking at History*, p. 24.
7. Louis A. Warren, *Lincoln's Youth*, pp. 3–13, 17 (map).
8. Danhof, *Change in Agriculture*, pp. 105–6. Admittedly those mortgages tended to run merely a few years' duration. Long-term mortgages remained exceptional until after the Civil War.
9. How people in-migrated from afar via the Ohio River was explained in detail in 1816 by French immigrant Marie Barbe Francois Lakanal. (See Michaux [ed.], "Instructions from the Ohio Valley to French Emigrants.") A topic-by-topic compilation of travel details drawn from numerous sources and covering all modes of travel is Jakle, *Images of the Ohio Valley*.
10. Esarey, *The Indiana Home*, pp. 11–17, 25–26; and Oliver Johnson, *A Home in the*

Woods, pp. xxiv–xxv, 6–12. Many additional details about Indiana's early settlement specifically from Kentucky, including the Ohio River crossing points and the travel routes up into Indiana, are in Bigham, *Towns and Villages of the Lower Ohio*, pp. 26–41. Gregory S. Rose shows that thousands of southern families used north-central Kentucky as a jumping-off point for their move into Indiana. Rose shows this by mapping the last place of previous residence reported by all the people who bought public land in Indiana prior to 1850. See Rose, "Upland Southerners," p. 254 map.

11. Rose, "Upland Southerners," pp. 259–60.
12. Several basic designs of log cabins and barns are illustrated in Sieber and Munson, *Looking at History*, pp. 37–41. Wild honey bee lore is especially well-explained by Buley, *The Old Northwest*, vol. 1, pp. 155–57.
13. Rohrbough, *The Trans-Appalachian Frontier*, pp. 17, 93–94; Esarey, *A History of Indiana*, vol. 2, pp. 580–82; Duncan, "Old Settlers"; and Carmony, *Indiana, 1816–1850*, pp. 68–69. Descriptions of how womenfolk went about performing these activities appear in Buley, *The Old Northwest*, vol. 1, pp. 202–25. On making lye and soap, see Esarey, *The Indiana Home*, pp. 28–29. Making buckskin, in turn, required both lye and soap. Among Yankee in-migrants, the cows were milked by men and boys rather than by women and girls. (Schob, *Hired Hands and Plowboys*, pp. 199–200.) But most early Hoosiers were not Yankees.
14. Esarey, *The Indiana Home*, p. 16. In all except the poorest families, the wife probably had a Dutch oven in addition to her three-legged skillet.
15. Ibid., pp. 15–16.
16. My own personal years of backwoods life in West Virginia in the 1970s included enough farm-making work to show me the accuracy of these three accounts.
17. Esarey, *The Indiana Home*, pp. 24–26; and Esarey, *A History of Indiana*, vol. 1, p. 421.
18. Duncan, "Old Settlers," pp. 395–96. Cabins could of course be built at any time of year, but the damp-clay "chinking" that was used to fill the cracks between the logs and slabs of wood was least likely to crack and fall out if the cabin was chinked in late fall, just before temperatures dropped below freezing. (Personal experience.)
19. Danhof, *Change in Agriculture*, pp. 90–92. The logic of thus starting out as tenants is set forth in Ankli, "Farm-Making Costs in the 1850s," pp. 58–59. But pitfalls that some Indiana tenants encountered are emphasized in a caution that was penned in 1856 and is quoted at length in Danhof, *Change in Agriculture*, pp. 93–94.
20. Although Midwestern Indians also used tree-deadening, the Euro-American settlers had been using that practice before they reached the Midwest, says Jones, *History of Agriculture in Ohio*, p. 10.
21. Esarey, *A History of Indiana*, vol. 1, pp. 425–26, and vol. 2, pp. 577–78 (which includes a description of how and when fence rails were split); Esarey, *The Indiana Home*, pp. 26, 76–77; Duncan, "Old Settlers," pp. 396–98; and Oliver Johnson,

A Home in the Woods, pp. 13–21. Regarding the convivial jug of whiskey, see Etcheson, *The Emerging Midwest*, p. 86.

22. Danhof, *Change in Agriculture*, p. 117.
23. The David W. Walton Blacksmith Shop Account Book, unpaged. Also see Esarey, *A History of Indiana*, vol. 1, p. 426. Illustrations of early shovel plows appear in McClelland, *Sowing Modernity*, pp. 113, 117–18. Several later upscale shovel plows and their rigs are illustrated in Hurt, *American Farm Tools*, pp. 10–11.
24. Samuel R. Brown, *The Western Gazatteer*, p. 149.
25. Duncan, "Old Settlers," pp. 384–86; Carmony, *Indiana, 1816–1850*, pp. 57–59; and Danhof, *Change in Agriculture*, pp. 175–76.
26. Latta, *Outline History of Indiana Agriculture*, p. 55.
27. Power, *Planting Corn Belt Culture*, p. 155; and Oliver Johnson, *A Home in the Woods*, p. xvi. The evolution of Indiana's hog markets and pork packers is summarized in Thornbrough, *Indiana in the Civil War Era*, pp. 418–19.
28. Buley, *The Old Northwest*, vol. 1, pp. 478–81; and Hudson, *Making the Corn Belt*, p. 80. Hudson adds that cattle likewise could be driven about ten miles a day, and that herds of hogs were often driven behind herds of cattle from Ohio to the East.
29. Oliver Johnson, *A Home in the Woods*, pp. xv–xvi; and Buley, *The Old Northwest*, vol. 1, pp. 169–70. To "bolt" meant the new grains of wheat appeared prematurely at the head of their stalks. Likewise on overly fertile ground, the stalks tended to grow too tall.
30. Esarey, *A History of Indiana*, vol. 1, p. 427; and Gates, *The Farmer's Age*, p. 170. Cradle scythes and how they were used are described and illustrated in Hurt, *American Farm Tools*, pp. 40–42; and in McClelland, *Sowing Modernity*, pp. 134–41.
31. Quoted in John E. Purple, "Greene County Agriculture," p. 1, in the Indiana Federal Writers' Project/Program Papers, 170 Agriculture, Indiana State University Library, Special Collections.
32. It *was* a waste of time, claimed Oliver H. Smith, in a letter of 28 April 1838 that is printed in Ellsworth, *Valley of the Upper Wabash*, pp. 39–41. On the other hand, such "hogging-off" or "hogging-down" of a field could waste a lot of corn, since hogs tended to be sloppy eaters.
33. Esarey, *A History of Indiana*, vol. 1, p. 428.
34. So says the naturalist Marion T. Jackson, personal communication in January 2004.
35. Lambert, *Traces of the Past*, pp. 216–17.
36. As late as 1842 in Indianapolis, saddles of fresh venison were selling for just 25 to 50 cents, and wild turkeys were selling for just 10 to 12 cents. Passenger pigeons sold for 25 cents a bushel. Buley, *The Old Northwest*, vol. 1, p. 153 n. 28.
37. Among many accounts, see Walsh, *The Rise of the Midwestern Meat Packing Industry*, pp. 18–19.

38. Buley, *The Old Northwest*, vol. 1, p. 154; and Nicholson, "Swine, Timber, and Tourism," p. 30.
39. Rutherford, *Wild Mustard*, pp. 70–71.
40. Cayton and Onuf, *The Midwest and the Nation*, p. 39; Esarey, *A History of Indiana*, vol. 1, pp. 269–72; and Sieber and Munson, *Looking at History*, pp. 27–28 including illustration. Besides flatboats, which were designed for downstream travel, keelboats were common. Keelboats were designed to be poled upstream against the current of rivers. Canal boats resembled keelboats. Details are in Jakle, *Images of the Ohio Valley*, pp. 27–31.
41. Esarey, *The Indiana Home*, p. 64; and Rohrbough, *The Trans-Appalachian Frontier*, p. 176.
42. Calculated from Esarey, *A History of Indiana*, vol. 1, p. 270. A decade later, in 1836, an estimated 800 flatboats would pass Vincennes en route to downriver markets. Note, however, that a significant portion of the Wabash River traffic originated on the Illinois side of the river. (Rohrbough, *The Trans-Appalachian Frontier*, p. 176.) A detailed map of Indiana's and eastern Illinois' rivers and streams is in Esarey, *A History of Indiana*, vol. 1, p. 267.
43. Gates, *The Farmer's Age*, p. 162. In fact, whiskey was Cincinnati's fourth leading export, by value, for the overall period up to the Civil War. Only pork, flour, and grain surpassed the value of Cincinnati's whiskey exports. Most of it, however, was not actually distilled in Cincinnati—although Ohio as a state did in 1850 lead all other states in the distilling of alcohol, with Indiana ranking fourth. Indiana held many distilleries. For instance, three that were located in southeastern Indiana's Dearborn County distilled $200,000 worth of whiskey in 1847–48, and probably sent much of that to Cincinnati. John G. Clark, *The Grain Trade in the Old Northwest*, pp. 17, 152, 224 n. 25.
44. Gates, *The Farmer's Age*, pp. 175–76; and John G. Clark, *The Grain Trade in the Old Northwest*, p. 25. Bulk pork was preserved either by heavily salting it and letting it lie in a cool cellar, or by both salting and smoking it. Then it was packed in 800 to 900 pound hogsheads to be shipped. Since such bulk pork was cheaper than barreled pork, it was often provided as rations to southern slaves. Pork that was packed in barrels (which weighed about 200 pounds when full) was packed in a heavy brine. Walsh, *The Rise of the Midwestern Meat Packing Industry*, pp. 32–33.
45. Rugh, *Our Common Country*, p. 22; and John G. Clark, *The Grain Trade in the Old Northwest*, pp. 24–25.
46. Lanier, *Sketch of the Life of J.F.D. Lanier*, p. 16. Such methods of financing midwestern trade were typical and are also described in Knodell, "The Demise of Central Banking and the Domestic Exchanges," pp. 715–16, 718; and in Berry, *Western Prices before 1861*, pp. 227–28. Lanier later became a key ally of Republican governor Oliver P. Morton, helping Morton to finance Indiana's participation in the Civil War.

47. Sauer, "Homestead and Community on the Middle Border," pp. 38–39.
48. Esarey, *The Indiana Home*, pp. 29–35. That the planting of potatoes was started on Good Friday is according to the naturalist Marion T. Jackson, personal communication in January 2004. An older source says the *flax* patch was supposed to be planted on Good Friday. (Buley, *The Old Northwest*, vol. 1, p. 203.) Flax was used to make linen cloth and also to interweave with wool into linsey-woolsey for everyday work clothes, especially women's. Duncan, "Old Settlers," pp. 389–91; and Carmony, *Indiana, 1816–1850*, pp. 53–54. Flax *seed* could be sold to pressers of linseed oil.
49. Legends about Johnny Appleseed are scrutinized systematically in Price, *Johnny Appleseed: Man and Myth.*
50. Mary Mace, letter of November 1820, described in Power, *Planting Corn Belt Culture*, p. 107.
51. Esarey, *The Indiana Home*, pp. 30, 33–35.
52. *History of Switzerland County, Indiana*, pp. 992–97.
53. Esarey, *A History of Indiana*, vol. 1, p. 271; and Dufour, *The Swiss Settlement of Switzerland County Indiana*, pp. xiii–22. Later generations of those Swiss families used the same land to grow more conventional crops.
54. Dean, *Journal of Thomas Dean*, pp. 45–46. Also see Sugden, *Tecumseh*, pp. 138–42, 164, 196, and especially p. 221.
55. Pitzer and Elliott, "New Harmony's First Utopians, 1814–1824," pp. 233–35, 238. For details, Arndt (ed.), *A Documentary History of the Indiana Decade of the Harmony Society*, vol. 1.
56. Butternuts' festivities are described in Duncan, "Old Settlers," p. 398.
57. Power, *Planting Cornbelt Culture*, pp. 2–3. Also see Mathews, *The Expansion of New England*, pp. 208–10 including map. Other assertions by historians who have discounted early Upland Southerners' enterprise are quoted in Hudson, *Making the Corn Belt*, pp. 92–93, 110–11.
58. Atack and Bateman, "Yankee Farming and Settlement in the Old Northwest," p. 85; also see p. 86 table.

Chapter Five

1. Hudson, *Making the Corn Belt*, pp. 66–74, 102–3. The label "corn belt" did not see print until about 1882 says Power, *Planting Corn Belt Culture*, p. 160.
2. On the evolution of *Zea mays* see Hudson, *Making the Corn Belt*, chapter 4; on extensive crop cultivation by Native Americans see Hudson's chapters 2 and 3; and on Euro-American explorers "spying out" the location of land already being used by Native Americans for intensive cropping, see Hudson's chapter 3.
3. Quoted in Hudson, *Making the Corn Belt*, pp. 67–68.
4. Hudson, *Making the Corn Belt*, chapter 1. Hudson's documentation as to who first brought the future corn-belt system westward is not new to agricultural

history. Parallel and additional documentation appeared in Paul C. Henlein's 1959 book *Cattle Kingdom in the Ohio Valley* (chapter 1); also in John Stealey's 1966 article "Notes on the Antebellum Cattle Industry"; and in a 1991 book chapter by Richard K. MacMaster, "The Cattle Trade in Western Virginia." At his page 128, MacMaster mentions still more sources.

5. MacMaster, "The Cattle Trade in Western Virginia," pp. 131–45 including tables; Hudson, *Making the Corn Belt*, pp. 67–69 including illustrations; and U.S. Census of 1860, *Agriculture in the United States in 1860*, pp. cxxx–cxxxiii. More details about the imported cattle are in Jones, *History of Agriculture in Ohio*, pp. 106–8.
6. Jones, *History of Agriculture in Ohio*, pp. 91–92. The use of the same basic feeding system in Illinois in the 1830s is documented by Whitaker, *Feedlot Empire*, pp. 24–29.
7. Berry, *Western Prices before 1861*, p. 218; and Jones, *History of Agriculture in Ohio*, pp. 80–82, 88–90.
8. Jones, *History of Agriculture in Ohio*, p. 90. Also see p. 86.
9. Hudson, *Making the Corn Belt*, pp. 96–97. Benjamin Franklin Harris had been born in the Shenandoah Valley in 1811 and had gone west to Ohio's Scioto Valley in 1831 with his father. In 1841 he relocated to land he had bought near Champaign, Illinois. By 1855 he was raising 700 acres of corn there and feeding it to 360 cattle and 200 hogs.
10. Hudson, *Making the Corn Belt*, pp. 10–12, 63–68, 88–109 (including maps on pp. 94, 100); and U.S. Census of 1860, *Agriculture in the United States in 1860*, p. cxxx.
11. Hudson, *Making the Corn Belt*, p. 103.
12. Hudson, *Making the Corn Belt*, pp. 108, 111.
13. Baker, "Indian Corn and Its Culture," pp. 96–97. Many purported genealogies of dent corn vary from this version. For instance, many Illinois farmers "claimed that the commonly used 'Little Yellow' had come from the Indians. In 1846 Robert Reid, of Ohio, moved to Tazewell County, Illinois, and planted Gordon Hopkins corn which he had brought along. When many hills failed to germinate, neighbors supplied Little Yellow for replanting. From this mixture eventually came Reid's Yellow Dent, a favorite of the corn country until the advent of hybrid corn." (Buley, *The Old Northwest*, vol. 1, p. 175 n. 77. Also see pp. 174–76 there.) Both flint and gourdseed ears are illustrated in Power, *Planting Corn Belt Culture*, opposite p. 139; and also in Hudson, *Making the Corn Belt*, pp. 51–52.
14. Hudson, *Making the Corn Belt*, pp. 7–8. "Improved" farmland is a standard census category and includes not only acreage under cultivation but acreage lying fallow and acreage used for pasture and orchards.
15. Hudson, *Making the Corn Belt*, pp. 136–38 including map. And in fact, as of 1955 cash-grain corn farming in *Indiana* remained most pronounced on its own slice

of the Grand Prairie—on the drained wetlands that lie just south of the Kankakee River in Newton, Jasper, Benton, and White counties. See the 1955 map of Indiana's regional "Farming Types" in Dillon and Lyon, *Indiana: Crossroads of America*, p. 79. Below in chapter 9 we'll look closer at that part of the state.

16. Hudson, *Making the Corn Belt*, pp. 8–9 including map.
17. Hudson, *Making the Corn Belt*, pp. 10–11 including map. During the later 1800s the corn belt kept creeping northward, but along its *southern* edge it began *withdrawing* northward because of soil depletion. See Hudson, pp. 151–57 including maps.
18. Whitaker, *Feedlot Empire*, p. 18. See also Hudson, *Making the Corn Belt*, p. 97. A detailed panorama of the westward shift of cattle fattening—from Virginia through Kentucky and Ohio and on to Indiana, Illinois, and points west—is provided by Whitaker, pp. 18–32.
19. Calculated from the U.S. Census of 1860, *Population of the United States in 1860*, p. 111, and *Agriculture of the United States in 1860*, p. 43.
20. As reported by the former commandant of Detroit, Jacques-Charles de Sabrevois. See Barnhart and Riker, *Indiana to 1816*, p. 67. Both the Wea and Piankashaws were branches of the Miami people.
21. Hudson, *Making the Corn Belt*, pp. 37, 88.
22. Sugden, *Tecumseh*, pp. 166–68.
23. Hudson, *Making the Corn Belt*, pp. 8–9 including map.
24. Phillips, *Indiana in Transition*, p. 148 for Indiana's 1909 corn acreage at 4,901,054 acres; and Thornbrough, *Indiana in the Civil War Era*, p. 365 for the total land surface of the state (excluding water surface) as being 21,637,760 acres. During World War I, Indiana's corn-planted acreage rose even higher, temporarily. In 1917, farm cornfields covered 5,875,000 Indiana acres, 27 percent of the state's land surface. Houk, "A Century of Indiana Farm Prices," p. 23.

Chapter Six

1. Cayton, *Frontier Indiana*, pp. 180, 265.
2. Gipson (ed.), *The Moravian Indian Mission on White River*, pp. 7, 8n, 9n, 60. Also see Cayton, *Frontier Indiana*, pp. 196–99, 262–63. A map of the Indian villages along White River appears in Thompson, *Sons of the Wilderness*, foldout between pp. 42–43; and reprinted in Crumrin, "Between Two Worlds," p. 20. A map of the boundaries of the 1818 New Purchase, and those of all other Indiana land cessions, can be found above on p. 61.
3. Both quotes are from Gipson (ed.), *The Moravian Indian Mission on White River*; see pp. 297, 450.
4. Crumrin, "Between Two Worlds." Also see Larson and Vandersteel, "Agent of Empire," and Thompson, *Sons of the Wilderness*, pp. 42–46. In 1934, Eli Lilly bought 1,400 acres of William Connor's old farm and financed the restoration of its mansion (built in 1823) and several farm buildings. In 1963, Eli

Lilly gave the farm to Earlham College to open to the public and today it is the Conner Prairie living history farm. See Stephen Cox, "New Life" and "Back to the 1820s."

5. Duncan, "Old Settlers," p. 396; also pp. 377–78. Plausibly William Conner's partner was the same William Marshall who had grown crops at Fort Wayne in 1804 as an employee of the Indian trader John Johnston. See Woehrmann, *At the Headwaters of the Maumee*, p. 95.
6. Duncan, "Old Settlers," pp. 383–88.
7. Duncan, "Old Settlers," pp. 378–79; Carmony, *Indiana, 1816–1850*, pp. 107–12; Esarey, *A History of Indiana*, Vol. 1, p. 343; and Rohrbough, *The Trans-Appalachian Frontier*, pp. 167–68, 176. Dietary options when corn ran short are mentioned in Buley, *The Old Northwest*, Vol. 1, pp. 154–55. A map showing several early mill locations just north of Indianpolis appears in Oliver Johnson, *A Home in the Woods*, p. vi. Ginette Aley shows the crucial role that mills played in pioneer-era Indiana. See Aley, "Grist, Grit, and Rural Society."
8. Schramm, *The Schramm Letters*, pp. 9–10, 45–50, 64–65.
9. Carmony, *Indiana, 1816–1850*, pp. 136–39.
10. Schramm, *The Schramm Letters*, pp. 45–47, 52–56, 67–68. Schramm claimed that Indianapolis bank clerks had never before seen gold coins and would deal only in banknotes. That's surely exaggerated, but Schramm did have to travel back to Cincinnati to exchange his *Austrian* gold coins for banknotes so he could buy the Boone County land.
11. Danhof, *Change in Agriculture*, p. 37. Factors that fostered (or delayed) early pork-packing enterprises are explained by Walsh, *The Rise of the Midwestern Meat Packing Industry*, pp. 7–14. One reason why *some* hogs were packed in barrels at home rather than driven to market was likely because they couldn't be captured, or be tamed enough to drive, so instead they were hunted and shot.
12. Schramm, *The Schramm Letters*, p. 68; and Gates, *Agriculture and the Civil War*, pp. 157–61.
13. Schramm, *The Schramm Letters*, pp. 82–83. That railroad was the Indianapolis, Winchester, & Bellefontaine, incorporated in 1848 and popularly known as "the Bellefontaine." When it was finished, it provided Indianapolis with a rail line through to Pittsburgh and the East. Meanwhile, also by the early 1850s, Indianapolis had become the terminus of seven railroads that brought produce there from other parts of the state. (John G. Clark, *The Grain Trade in the Old Northwest*, p. 150 including n. 8.) The dates of Indiana's early railroad incorporations appear in Morrison, *Indiana: "Hoosier State,"* Vol. 1, p. 28. Details about many early railroad lines, including the Bellefontaine, appear in Thornbrough, *Indiana in the Civil War Era*, pp. 322–46. As regards problems that farmers faced with sheep, see Danhof, *Change in Agriculture*, p. 153.
14. Schramm, *The Schramm Letters*, pp. 82–83. Indiana state bonds had earlier gone belly up after the financial panic of 1837—which had hit just one year after state

bonds were issued extravagantly to finance an array of canals and turnpikes, notably the Wabash & Eire Canal. Details of Indiana's post–1837 financial default are in Carmony, *Indiana, 1816–1850*, pp. 204–45, 293–312.

15. Schramm, *The Schramm Letters*, pp. 83–84. More will be said in chapter 9 about burying cylindrical tile drains in wet fields to expedite water runoff.
16. Nation, *At Home in the Hoosier Hills*, pp. 33–36; Rugh, *Our Common Country*, pp. 16–17, 65–67; Danbom, *Born in the Country*, pp. 87–90. A very detailed argument that early farm women's work was indispensable is set forth by Faragher, "The Midwestern Farming Family, 1850."
17. In southern Indiana, cotton too was spun and woven at home by women. First "the seeds were picked or combed out of the boll by the hands of the pioneer women after which the cotton was carded and spun into strands from which garments and other cloth items were fabricated." John E. Purple, "Greene County Agriculture," p. 1, in the Indiana Federal Writers' Project/Program Papers, 170 Agriculture, Indiana State University Library, Special Collections.
18. Foster, ed., *American Grit*, pp. 33–34, 35–36.
19. Carmony, *Indiana, 1816–1850*, pp. 77–78; and Carter, "Rural Indiana in Transition," pp. 114–15. How hogs were butchered, and apple butter and maple sugar were made, is detailed in Buley, *The Old Northwest*, Vol. 1, pp. 213–16, 219–20, 224–25. The average prices fetched by eggs each year are set forth by Houk, "A Century of Indiana Farm Prices," pp. 52–53 table.
20. U.S. Census of 1860, *Population of the United States in 1860*, p. 111. How midwestern farm women were faring fifty years later, during the early 1900s, is examined by Neth, *Preserving the Family Farm*, chapter 1, which also documents the early 1900s' division of midwestern farm labor among men, women, and children.
21. Buley, *The Old Northwest*, Vol. 1, p. 309.
22. Carmony, *Indiana, 1816–1850*, pp. 78–79. Precise recommendations of which chores should be shifted from women to men, along with the reasons why, appeared in the first annual report of the U.S. Dept. of Agriculture (in 1862) and are summarized in Faragher, "The Midwestern Farming Family, 1850," pp. 125–26.
23. Fuller, *Summer on the Lakes*, p. 61.
24. Schob, *Hired Hands and Plowboys*, pp. 191–92.
25. Modell, "Family and Fertility on the Indiana Frontier, 1820," p. 616. Figures there for Ohio and Pennsylvania show a similar decline from 1820 to 1830 and 1840, falling in Ohio from 2.13 to 1.87 to 1.7, and in Pennsylvania from 1.75 to 1.54 to 1.46.
26. John Johnston, in Hill, *John Johnston and the Indians*, p. 191.
27. Cayton and Onuf, *The Midwest and the Nation*, pp. 46–47.
28. Fuller, *Summer on the Lakes*, p. 61.
29. Cayton and Onuf, *The Midwest and the Nation*, p. 46. Susan Sessions Rugh shows in her book *Our Common Country* that, in west-central Illinois, not only Yankees and Upland Southerners tended to settle in clusters but that Pennsyl-

vanians did too. (Rugh, pp. 10–16.) Part of those Pennsylvanians' motive for clustering may have been to swap labor with each other during wheat harvesting and threshing. (Rugh, pp. 14–20.) Immigrants from Europe also tended to cluster together if they came from the same country.

30. Sturm, "The Frontier," p. 266.
31. Schwartzweller, Brown, and Mangalam, *Mountain Familes in Transition*, pp. 90–98.
32. A detailed chronology of one *English*-American family's successive kin-mediated moves from coastal Maryland through Virginia, North Carolina, and Kentucky before settling in south-central Indiana's hilly Brown County appears in Nicholson, "Swine, Timber, and Tourism," pp. 11–22. That family apparently moved into Brown County's hills to distance itself from "the ague" (malarial fever) and from landlords.
33. Vincent, *Southern Seed, Northern Soil*, pp. xv–3 (including maps and table), 26–29, 38–44, 62. More details are in Cord, "Black Rural Settlements in Indiana before 1860," as too is information about African-American farm families who had migrated earlier to *southern* Indiana. Cord maps twenty-one pre-Civil War rural African-American settlements in Indiana (p. 101). Likewise, Euro-American families also tended to settle near other families from their own place of origin—at least until commercial "railroad farming" proliferated in the 1850s. Cayton and Onuf, *The Midwest and the Nation*, pp. 39–46. Examples are in Rohrbough, *The Trans-Appalachian Frontier*, pp. 140–42.
34. The origins and aims of the new rural history are told in Swierenga, "The New Rural History," and Swierenga, "Theoretical Perspectives on the New Rural History." Many examples of such studies are mentioned in Rothstein, *Writing American Agricultural History*, pp. 18–19, 47–48 (notes 46–48).
35. Vincent, *Southern Seed, Northern Soil*, pp. 39–40. A similar case of stem-family migration brought a number of African Americans to Brown County, as recounted in Nicholson, "Swine, Timber, and Tourism," pp. 94–96.
36. Quakers bought land in Rush County partly because it was close to Wayne County, whose county seat of Richmond was the site of a Quaker Yearly Meeting. Many North Carolina African Americans had accompanied Quakers to Wayne County, and as of 1850 its 1,036 African Americans constituted almost one-tenth of Indiana's total African-American population. Quite a few of Wayne County's African Americans lived near the Underground Railroad "station" of the Quaker abolitionist Levi Coffin. (Cord, "Black Rural Settlements in Indiana before 1860," p. 105; Vincent, *Southern Seed, Northern Soil*, pp. 35–38; and U.S. Census of 1850, *Statistical View of the United States*, pp. 63, 230.) An overview of antebellum African-American migrants from the South to Indiana, and of their relations with Quakers, is Thornbrough, *The Negro in Indiana*, chapter 2.
37. Vincent, *Southern Seed, Northern Soil*, pp. 40–43 (including map), 57–62 (including tables).

38. Cord, "Black Rural Settlements in Indiana before 1860," pp. 102, 106.
39. See, e.g., Ellsworth, *Valley of the Upper Wabash*, pp. 5, 68–77, 102–7, 149–73.
40. William Henry Smith, *The History of the State of Indiana*, p. 355.
41. Population growth calculated from Prince, *Wetlands of the American Midwest*, p. 149 table.

Chapter Seven

1. Esarey, *A History of Indiana*, vol. 1, pp. 273–78.
2. Morrison, *Indiana: "Hoosier State,"* vol. 1, p. 12. Esarey dates the opening of the Crawfordsville land office for full-time business as occurring between one and five years later, by 1828 at the latest. Esarey, *A History of Indiana*, vol. 1, p. 344.
3. Atack and Bateman, *To Their Own Soil*, pp. 77–79; and Parker, *Europe, America, and the Wider World*, vol. 2, p. 140.
4. Esarey, *A History of Indiana*, vol. 1, pp. 276–78 (including 1830 population numbers county-by-county), and pp. 344–51 (including table of yearly land sales and payments received).
5. Schramm, *The Schramm Letters*, pp. 53–54. Each government land auction opened up for sale a new block of surveyed public land. Typically, an auction lasted several weeks.
6. Esarey, *A History of Indiana*, vol. 1, pp. 344, 350 table.
7. Billington, *Westward Expansion*, p. 321; and Oberly, *Sixty Million Acres*, p. 15 table.
8. Adams, "The Role of Banks in the Economic Development of the Old Northwest," p. 231; and Carmony, *Indiana, 1816–1850*, pp. 559, 832 n. 184.
9. For an appraisal of this long-standing debate, see Ankli, "Farm-Making Costs in the 1850s." Evidence exists for both sides of the debate, but, in Indiana's case, far more settlers arrived during economic swells than during hard times.
10. Public land was almost always first "opened" for sale at a pre-advertised land auction where the highest bidder got the land. Much of the public land didn't find buyers at the auctions, and then it stayed "open" for sale at the minimum price of $2 an acre (which after 1820 became $1.25 an acre).
11. North, *Growth and Welfare in the American Past*, pp. 79–81 (including charts), 93, 126–27.
12. Easterlin, "Farm Production and Income in Old and New Areas at Mid-Century," p. 83 map.
13. See Atack and Bateman, *To Their Own Soil*, pp. 38–40 including graph. In 1860, children aged nine or less made up almost one-third of the rural midwestern population, but only 0.3 percent of rural Midwesterners were aged 75 or older.
14. Rohrbough, *The Trans-Appalachian Frontier*, pp. 168–69, 173. Another 1820s memoir of farming in Parke County is Carmony, "From Lycoming County, Pennsylvania, to Parke County, Indiana."
15. A map of "Major Transportation Routes to Indiana during the Pioneer Period"

(including five roads, six rivers, and four canals) appears in Rose, "Hoosier Origins," p. 227.

16. The David W. Walton Blacksmith Shop Account Book, unpaged. Note that "sang hoes," besides their use for digging up ginseng roots, were also handy for planting corn in rough fields that were newly slashed and burned. Vance, *Human Geography of the South*, p. 415.
17. Schramm, *The Schramm Letters*, p. 76. These were prices for ordinary scythes, not for cradle scythes, which cost several times more. Prices a bit later (in the 1850s) of innumerable other items, including dry goods, household furnishings, food, and personal accessories, can be found in Carter, "Rural Indiana in Transition," pp. 116–17.
18. The David W. Walton Blacksmith Shop Account Book, unpaged. Another blacksmith's account book, dating a couple years later from the Sangamon Valley in central Illinois, suggests that *most* exchanges between farmers and blacksmiths entailed little or no use of cash. See Faragher, "Open-Country Community," pp. 245–46.
19. The David W. Walton Blacksmith Shop Account Book, unpaged.
20. Solon Robinson, in Kellar (ed.), *Solon Robinson*, vol. 1, p. 261. These prices were as of 1841 in northwestern Indiana.
21. Solon Robinson, in Kellar (ed.) *Solon Robinson*, vol. 1, pp. 279, 294.
22. Power, *Planting Corn Belt Culture*, p. 155.
23. Oliver H. Smith, letter of 23 April 1838, printed in Ellsworth, *Valley of the Upper Wabash*, p. 41.
24. Buley, *The Old Northwest*, vol. 1, pp. 480–81.
25. Fatout, *Indiana Canals*, p. 39. Details about the origins of the Wabash & Erie canal idea appear in Benton, *The Wabash Trade Route*, pp. 32–57. The canal idea harmed the Miami Indians by hastening their forced removal from northern Indiana to Indian Territory west of the Mississippi River. Carmony, *Indiana, 1816–1850*, pp. 496–98.
26. Esarey, *A History of Indiana*, vol. 1, pp. 351–57 including map.
27. Howe, *A Descriptive Catalogue of the Official Publications*, pp. 190–92; Fatout, *Indiana Canals*, pp. 72–73; and Benton, *The Wabash Trade Route*, pp. 50–56.
28. Among many accounts, Buley, *The Old Northwest*, vol. 2, pp. 280–87 is mercifully succinct. Both self-serving mendacity and some officials' incompetence contributed to the sorry mess.
29. Fatout, *Indiana Canals*, p. 103; and Benton, *The Wabash Trade Route*, p. 109. By 1851, over 88,000 barrels of salt were going west each year on the canal, mainly for use in packing pork (Benton, p. 102). The boat trip between Lafayette and Toledo took about three days and nights. Packet boats for passengers took somewhat less than that, whereas commodity shipments often took longer.
30. Esarey, *A History of Indiana*, vol. 1, pp. 276–78; and Benton, *The Wabash Trade*

Route, pp. 93–102. Benton notes that trade routes such as the Wabash & Erie Canal drew a higher percentage of foreign immigrants among their newcomers than did other areas, perhaps partly because foreign immigrants generally lacked wagons, etc. and thus had to travel by public transportation. Along the Wabash & Erie, most foreigners were either Irish or German, the Irish having come mainly as canal-digging laborers. Benton, pp. 96–98.

31. Benton, *The Wabash Trade Route*, p. 101.
32. Benton, *The Wabash Trade Route*, pp. 87, 102–4.
33. Further possibilities among corn-hog options are spelled out by Walsh, *The Rise of the Midwestern Meat Packing Industry*, pp. 7–11, 18–24.
34. Sauer, "Homestead and Community on the Middle Border," p. 38. Also see Carter, "Rural Indiana in Transition," p. 108.
35. Faragher, "Open-Country Community," pp. 245–47; Faragher, "Americans, Mexicans, Métis," pp. 97–99; Faragher, *Sugar Creek*, pp. 133–34; and Cayton and Onuf, *The Midwest and the Nation*, pp. 32–34.
36. An economic analysis of why this was true is in Post, "The 'Agricultural Revolution' in the United States," pp. 218–19.
37. Goodman, "The Emergence of Homestead Exemption in the United States," pp. 470, 472 (table), 476; and Saynor, *The Development of Southern Sectionalism*, pp. 118–20. Regarding the federal debt relief for land purchasers, see Carmony, *Indiana, 1816–1850*, pp. 462–63; and Rohrbough, *The Land Office Business*, chapter 7. Later, the financial crash of 1873 led to another long depression (the third long depression of the 1800s' total of four) and in 1878 the Indiana General Assembly raised the homestead exemption to $600 and added further protections. Thornbrough, *Indiana in the Civil War Era*, p. 315.
38. Easterlin, "Farm Production and Income in Old and New Areas at Mid-Century," pp. 82–85 (including map), 100–3.
39. U.S. Census of 1860, *Agriculture of the United States in 1860*, p. clxvii. An average hog weighed about 200 to 250 pounds when sold, and it took about fifteen bushels of corn to fatten one hog for market. (Rastatter, "Nineteenth Century Public Land Policy," pp. 126–27.) The prices of corn and hogs rose and fell roughly in tandem with each other. Shannon, *The Farmer's Last Frontier*, pp. 165–68 including graph.
40. Berry, *Western Prices Before 1861*, p. 126.
41. Earle, *Geographical Inquiry and American Historical Problems*, p. 312.
42. Hughes, *American Economic History*, p. 185.
43. Headlee, *The Political Economy of the Family Farm*, p. 33.
44. U.S. Census of 1860, *Agriculture of the United States in 1860*, p. clxvii.
45. Solon Robinson's veritable sermons in this vein enliven his 1840s agricultural writings. See for example Solon Robinson in Kellar (ed.), *Solon Robinson*, vol. 1, pp. 143, 147, 263, 343–56, 367–71.

Chapter Eight

1. Power, *Planting Corn Belt Culture*, p. 111.
2. Crop of 1839 as reported in the U.S. Census of 1840, *Compendium*, pp. 287–88. For production details township-by-township, see the *Statistics* volume of that census, pp. 328–41. Township boundaries as of 1876 are mapped in the *Illustrated Historical Atlas of the State of Indiana*. Today's Indiana township boundaries are mapped in Andriot (compiler), *Township Atlas of the United States*, pp. 256, 269–77.
3. Solon Robinson, in Kellar (ed.), *Solon Robinson*, vol. 1, pp. 139–40, 144–46, 240. That failure of the 1840 wheat crop did not distort the 1840 U.S. agricultural census since the crop yields recorded in the Census were always those of the pre-census year, in this case 1839's. Note, however, that livestock counts were officially made as of January 1 of the census year itself (January 1, 1840 in this case).
4. Gates, *The Farmer's Age*, pp. 163–64, 167–68. One early student of the Hessian fly as a threat to wheat was Thomas Jefferson. When Alexander Hamilton issued his famous *Report on Manufactures* in 1791, Jefferson started two urgent projects which both took him to New York State. One was to devise a political faction to rein in Hamilton and the other was to study the Hessian fly—the latter so that the U.S. could prosper financially through agriculture and not need to make its own manufactured goods. Jefferson devised some fly strategies that helped, and wrote them up for the American Philosophical Society. Both of Jefferson's 1791 New York projects are described in Randall and Nahra, *American Lives*, vol. 1, pp. 125–32.
5. Solon Robinson, in Kellar (ed.), *Solon Robinson*, vol. 1, pp. 120–21, 287–88; and Ball, *Northwestern Indiana*, pp. 90–92. Most accounts speak of plow cuts only two to four inches deep, and as one author put it, "a year or more was required for the turf to rot." Buley, *The Old Northwest*, vol. 1, p. 175.
6. Hurt, *American Farm Tools*, pp. 12–18; Thornbrough, *Indiana in the Civil War Era*, p. 375; and Furlong, "Plowmakers for the World."
7. Solon Robinson, in Kellar (ed.), *Solon Robinson*, vol. 1, pp. 120–21, 262, 368. The 1850 *Indiana Gazetteer* confirmed, regarding corn, that "a single hand can prepare the ground, plant, attend to, and gather, from 20 to 25 acres." Quoted in Carter, "Rural Indiana in Transition," p. 115.
8. How wheat was harvested with scythes and threshed with flails prior to horse-drawn and horse-turned mechanization is summarized by William Henry Smith, *The History of the State of Indiana*, pp. 252–53. On mechanical reapers' daily acreage of harvest, see Thornbrough, *Indiana in the Civil War Era*, p. 379. Illustrations and mechanical explanations of reapers appear in Hurt, *American Farm Tools*, chapter 5. There, "self-raking" reapers are illustrated and explained on pp. 46–49. The Civil War era's manpower shortage would make self-raking reapers highly popular.

9. Atack and Bateman, *To Their Own Soil*, pp. 194–200; Headlee, *The Political Economy of the Family Farm*, p. 63; and David, *Technical Choice, Innovation and Economic Growth*, p. 199 including n. 3. *How* reapers were adapted to do double duty as hay mowers is explained by Jones, *History of Agriculture in Ohio*, pp. 274–76.
10. Elsmere, *Henry Ward Beecher*, p. 200.
11. Rikoon, *Threshing in the Midwest*, chapters 1 and 2, including p. 5 photo of Herbert Kleinman flailing wheat in west-central Indiana's Parke County in 1981. After the threshing, there were even more choices among ways to winnow the wheat and thereby remove the chaff.
12. On early mobile steam-powered threshers, see Solon Robinson, in Kellar (ed.), *Solon Robinson*, vol. 1, pp. 240–41, 322, 368. Thornbrough emphasizes that most wheat farmers didn't acquire their own threshing machines because they cost many hundreds of dollars. Instead, they hired the machine and its operator. But most wheat farmers *did* eventually acquire their own reaper. (Thornbrough, *Indiana in the Civil War Era*, pp. 374–79.) Illustrations and mechanical details of threshers appear in Hurt, *American Farm Tools*, chapter 7. Jones, *History of Agriculture in Ohio*, pp. 270–72 explains the steps through which threshers evolved from the 1820s to the 1880s. Jones says that the reason why threshers were sometimes put on wheels or on wagons and pulled through the fields, threshing as they went, was to get wheat to market as soon as possible—before its selling price went down. But he asserts that field-threshing was not started in Ohio until the mid 1850s.
13. Rikoon, *Threshing in the Midwest*, chapter 3.
14. Alice Demmon, "Folklore: Harvest Time," pp. 1–3, in the Indiana Federal Writers' Project/Program Papers, Lake County, 170 Agriculture, Indiana State University Library, Special Collections. Also see Rikoon, *Threshing in the Midwest*, chapter 6.
15. Wallace Brown, "A Typical Threshing Day of the Yesterday," in the Indiana Federal Writers' Project/Program Papers, Shelby County, 170 Agriculture, Indiana State University Library, Special Collections. The meals and culture of the threshing season are described in even more detail by Rikoon, *Threshing in the Midwest*, chapter 6.
16. Rikoon, *Threshing in the Midwest*, chapter 4; and Gates, *Agriculture and the Civil War*, pp. 237–38. The steam-powered threshers cost $1,500 to $4,000 so *very* few of their owners were farmers. They could thresh up to 100 bushels of grain per hour. A photo of such a thresher in operation appears in Neth, *Preserving the Family Farm*, p. 45.
17. Douglas Harper, *Changing Works*, pp. 160–61. "Changing works" is what upstate New York farmers called their pervasive exchange of work with each other—their neighborly reciprocity.
18. Wayne Price, "Decatur County Wheat Crop . . . ," pp. 1–2, in the Indiana Federal

Writers' Project/Program Papers, Decatur County, 170 Agriculture, Indiana State University Library, Special Collections. A socioeconomic perspective on such "threshing rings" (as they were also called) is in Nicholson, "Swine, Timber, and Tourism," pp. 127–30; and in Rikoon, *Threshing in the Midwest*, chapter 5.

19. Traditional work exchanges in threshing and other farm tasks in upstate New York are minutely described and profusely illustrated as of the early to mid–1900s in Douglas Harper, *Changing Works*. On neighborly cooperation during threshing, see pages 113–36, 159–81.
20. Solon Robinson, in Kellar (ed.), *Solon Robinson*, vol. 1, pp. 322–24, 344–46, 378–80, 393–94. Monthly Indiana prices for wheat, reaching back to August 1841, appear in Houk, "A Century of Indiana Farm Prices," pp. 61–62 table. "Superior" breeds of hogs became widespread on Indiana farms in the 1840s. See Walsh, *The Rise of the Midwestern Meat Packing Industry*, p. 19.
21. Solon Robinson, in Kellar (ed.), *Solon Robinson*, vol. 1, p. 374.
22. Solon Robinson in the *Prairie Farmer*, September 1850, pp. 278–79, quoted in Whitaker, *Feedlot Empire*, p. 20. The advantages of producing corn and livestock in the geographical corn belt, rather than committing heavily to wheat, are spelled out in greater detail in Rugh, *Our Common Country*, pp. 60–61, 63–64.
23. Gates, *The Farmer's Age*, p. 171.
24. U.S. Census of 1860, *Agriculture of the United States in 1860*, pp. 38–39, 42–43. For Indiana's wheat-growing distribution as of 1879, Thornbrough, *Indiana in the Civil War Era*, p. 372. For 1899's wheat distribution, Higgs, *The Transformation of the American Economy*, p. 83 map. Wisely, southwestern Indiana farmers retained their crop flexibility—partly by starting in 1875 to grow muskmelons in some of their sandiest soils. About 1900 they switched to cantaloupe and found great success. Watermelons and sweet potatoes have also thrived in sandy soils there. (Donald Moore, "Sand Farming in Gibson County," pp. 1–3, in the Indiana Federal Writers' Project/Program Papers, Gibson County, 170 Agriculture, Indiana State University Library, Special Collections.) The 1860 census figures make Gibson County's neighbor Pike County appear the liveliest site of early Indiana melon-growing. U.S. Census of 1860, *Agriculture of the United States in 1860*, pp. 40, 44.
25. Parker, *Europe, America, and the Wider World*, vol. 2, p. 141.
26. Horace Greeley, quoted in Chandler, *Land Title Origins*, p. 491; and see pp. 489–92 there.
27. John Ade, *Newton County*, p. 2. At that time (in 1849) with a total population just shy of 30,000, one study has found that 75 percent of Chicago's adult males were living from hand to mouth, and that over 50 percent of Chicago's wealth was held by one percent of its adult males. See Cayton and Onuf, *The Midwest and the Nation*, p. 138 n. 1.
28. Walsh, *The Rise of the Midwestern Meat Packing Industry*, pp. 20–21 table, 50–54; Hudson, *Making the Corn Belt*, pp. 96–101, 107 (map), 130–32; Cronon, *Nature's*

Metropolis, pp. 26–30, 53–54; and U.S. Census of 1860, *Agriculture of the United States in 1860*, p. cxlix. The history of Chicago's stockyards is summarized in Whitaker, *Feedlot Empire*, pp. 41–44.

29. Adjustments that some farm families had to make after they moved to Chicago in the late 1800s are described in Sennett, *Families Against the City*, pp. 141–54, 173–237. Insights about the typical stages of a rural-to-urban transition appear in Bushman, "Family Security in the Transition from Farm to City."
30. Thornbrough, *Indiana in the Civil War Era*, p. 385. Also see Ball, *Northwestern Indiana*, pp. 408–9. Lake County's number of dairy cows as of 1870 and 1880 was exceeded only by Allen County, which contains Fort Wayne and is also *larger* than any other Indiana county. Later, far southeastern Indiana became part of Cincinnati's "milkshed." See the map of Indiana's twelve regional "Farming Types" as of 1955 in Dillon and Lyon (eds.), *Indiana: Crossroads of America*, p. 79.
31. "Fifty Year Survey of Dairying [in Indiana from 1870 to 1923]," p. 2, in Papers of Gov. James P. Goodrich, Herbert Hoover Presidential Library, West Branch, Iowa, Box 1, file "1927–35 and undated." This economic report on Indiana's dairy transition is part of a large batch of correspondence and data, both Indianan and national, that ex-governor Goodrich assembled in the mid–1920s to inform himself. His basic conclusion was that (as he phrased it in a 1927 letter) "I don't think [the farm situation] is one-half as bad as painted." James P. Goodrich to G. B. Hippie, 1 April 1927; in said Papers, Box 1, file "Agriculture: Foreign Trade Export Association Speech, 1927."
32. "Fifty Year Survey of Dairying," in ibid., Box 1, file "1927–1935 and undated," p. 1. By the years of Goodrich's governorship (1917–1921), the northwest corner's muck lands would be producing almost 60 percent of commercial U.S. peppermint, and also a significant slice of its commercial onions. (Phillips, *Indiana in Transition*, pp. 154–55.) Many other crops that thrive in muck soils are listed in Ethel Carmichael, "Agriculture (Kosciusko County)," in the Indiana Federal Writers' Project/Program Papers, Kosciusko County, 170 Agriculture, Indiana State University Library, Special Collections.
33. Carmony, *Indiana, 1816–1850*, pp. 137–39; and Esarey, *A History of Indiana*, vol. 1, pp. 257–60. The vast Kankakee Marsh continued another fifty years as a trappers' and hunters' paradise. Reputedly it sheltered more species of game birds than any other location in the United States. Carter, "Rural Indiana in Transition," p. 110. Also see George Ade, "Prairie Kings of Yesterday," p. 77; and Blakey, *Creating a Hoosier Self-Portrait*, pp. 159–64. Below in the next chapter we'll glance at this vast marsh as it began to be drained a bit later.
34. Harriet Martineau, 1836, in McCord (compiler), *Travel Accounts of Indiana*, pp. 161–62.
35. Ball, *Northwestern Indiana*, pp. 347–48; and Benton, *The Wabash Trade Route*, p. 109.

36. John Ade, *Newton County*, p. 103.
37. Hudson, *Making the Corn Belt*, pp. 96–100 including maps; Gates, "Hoosier Cattle Kings," pp. 1–5, 21–23; and Gates, *Landlords and Tenants on the Prairier Frontier*, pp. 108–39. Many of the names, counties, acreages, and idiosyncrasies of those land barons appear in George Ade, "Prairie Kings of Yesterday," especially at pp. 76–77 there.
38. Gates, *Landlords and Tenants on the Prairie Frontier*, pp. 72–107; Rose, "Upland Southerners," pp. 262–63; and Prince, *Wetlands of the American Midwest*, p. 164. Prince shows that as of 1870, over half of the Indiana farms which exceeded 1,000 acres in size were located in the state's northwestern Grand Prairie counties. See p. 187 map.
39. Henlein, *Cattle Kingdom in the Ohio Valley*, pp. 18–19; and Whitaker, *Feedlot Empire*, p. 116. Such prairies generally could not support cattle over the winter, however, so the cattle had to be moved elsewhere until spring returned.
40. George Ade, "Prairie Kings of Yesterday," p. 77; Jones, *History of Agriculture in Ohio*, pp. 86, 90; and U.S. Census of 1860, *Agriculture of the United States in 1860*, p. cxxxi. The size of Edward Sumners' holdings was 30,000 acres. (Gates, *The Farmer's Age*, p. 193.) The railroad that bisected Benton County in the 1850s was the Big Four Railway. Later it was taken over in 1914 by the New York Central.
41. Gates, *Agriculture and the Civil War*, p. 180. Solon Robinson inveighed against denying water to the cattle en route. Gates does not specify whether that denial of water was an oversight or a policy of the railroads.
42. Hudson, *Making the Corn Belt*, pp. 80, 99, 131–34; Prince, *Wetlands of the American Midwest*, pp. 203–36 (including p. 232 map); Whitaker, *Feedlot Empire*, pp. 77–78; and Anderson, *Refrigeration in America*, pp. 50–52.

Chapter Nine

1. Lloyd E. Cutler, "The Agricultural History of Lake County," p. 2, in the Indiana Federal Writers' Project/Program Papers, Lake County, 170 Agriculture, Indiana State University Library, Special Collections.
2. Woods, *The First Hundred Years of Lake County, Indiana*, p. 18.
3. Jakle, *Images of the Ohio Valley*, p. 46; and Gates, *The Farmer's Age*, pp. 181–82. The exertions of travel enhanced malaria's misery. Travelers' ordeals are described by Dean, *Journal of Thomas Dean*, pp. 63–73.
4. Duncan, "Old Settlers," pp. 400–2. On the prevalence of malaria in southwestern Indiana in the 1820s, see Miller, "Doctors, Drugs, and Disease in Pioneer Princeton," pp. 141–48. Native Americans too fell victim to malaria, as recounted by John Johnston in Hill, *John Johnston and the Indians*, p. 22.
5. Roberts, *Autobiography of a Farm Boy*, pp. 65–70.
6. Kellar, in Kellar (ed.), *Solon Robinson*, vol. 1, pp. 12–13. Also see pp. 68–76. Later in the same year (1839) the La Porte land office was moved from La Porte to Winamac.

7. Nationally authorized squatters' rights had begun in 1830. Through the Pre-Emption Acts of 1830, 1832, and 1834, Congress had legalized squatting on public land. Squatters' rights were then abolished by Congress in 1837 (and thus they were not in effect when the 1839 La Porte auction occurred). Congress restarted them again in 1841 and strengthened them thereafter, culminating in the 1862 Homestead Act. Between 1841 and 1862, whoever squatted on "pre-opened" land had to merely pay the minimum price of $1.25 per acre for it when it came up for auction sale. Otherwise the "preemption" was cancelled and ownership of that parcel went to its highest bidder—or if the parcel remained unsold after the auction, ownership went to whoever paid $1.25 an acre for it first. See Rohrbough, *The Land Office Business*, chapter 10; and Prince, *Wetlands of the American Midwest*, pp. 155–56.
8. See Esarey, *A History of Indiana*, vol. 1, pp. 349–50. How squatters' unions thwarted speculators is explained by Carmony, *Indiana, 1816–1850*, p. 461; by Gates, *Landlords and Tenants on the Prairie Frontier*, pp. 110–12; and by Buley, *The Old Northwest*, vol. 2, pp. 150–53.
9. Kellar, in Kellar (ed.), *Solon Robinson*, vol. 1, pp. 12–14.
10. Carmony, *Indiana, 1816–1850*, pp. 63–66. Most of those county-level agricultural societies disbanded or atrophied after the financial crash of 1837, but later many of them revived during the mid-1850s' boom.
11. Kellar, in Kellar (ed.), *Solon Robinson*, vol. 1, pp. 3–20. The Potawatomis' extensive commercial activities in northwestern Indiana and northeastern Illinois were centered on Fort Dearborn and its adjacent settlement, Chicago. (See Cronon, *Nature's Metropolis*, pp. 26–29, 53–54.) The principal year of Potawatomi expulsion was 1838 and the first stages of their sad journey that year were illustrated and described by the artist George Winter. See Winter, *Indians and a Changing Frontier*, pp. 99–107. The expulsions took several years. A Boston magazine reported that 1840's emigrating Potawatamis were supplied by government contractors with boxes that were "marked with large sums on the outside, but when opened were found to contain not one tenth part of the value." ("Frauds Upon the Indiana," *Boston Weekly Magazine* 3 [1841], p. 230, quoted in Banner, *How the Indians Lost Their Land*, p. 225.) Stuart Banner comments that "there had always been government contractors willing to exploit the Indians' lack of equal access to the legal system within which promises were enforced" (p. 225).
12. Kellar, in Kellar (ed.), *Solon Robinson*, vol. 1, pp. 16–38. The sheep fad was dealt a blow by the 1837 financial crisis but it peaked again in the 1850s. Solon Robinson's franking privilege as a postmaster saved the distant recipients of his agricultural letters from having to pay the high postage rates of those days. (See ibid., pp. 392–93 n. 1; and McCutcheon, *Everyday Life in the 1800s*, p. 84.) Robinson's first farmhouse design appeared in 1847 and emphasized a large kitchen. The layout is reproduced in McMurry, "Progressive Farm Families and

Their Houses," pp. 342–43. Barn architecture interested Robinson as well (see Kellar [ed.], vol. 1, pp. 436–37). Photos of typical early Indiana barns appear in Scott, *Barns of Indiana*, pp. 28–29, 89, 108, 118, 147.

13. Shannon, *The Farmer's Last Frontier*, p. 168. Also see Thornbrough, *Indiana in the Civil War Era*, p. 382; and Danhof, *Change in Agriculture*, p. 178. More details about the origins of both Berkshire and China-Poland hogs appear in Buley, *The Old Northwest*, vol. 1, pp. 188–89 including n. 129. For Solon Robinson's misadventures with razorback hogs, see his tirade in Kellar (ed.), *Solon Robinson*, vol. 1, pp. 129–31.
14. Solon Robinson, in Kellar (ed.), *Solon Robinson*, vol. 1, pp. 288–89, 320–21.
15. For instance, Henry William Ellsworth in his 1838 book *Valley of the Upper Wabash*, p. 35.
16. Gates, *The Farmer's Age*, pp. 187–88. Another fad fraught with unintended consequences was the government's early 1900s' promotion of the multiflora rose as a fencerow plant to provide nesting cover for game birds like pheasants and quail, and also to shelter rabbits. Multiflora rosebushes soon took over many pastures and rendered them useless, but the USDA kept recommending their propagation as late as 1967—as in U.S. Dept. of Agriculture, *Soil Survey: Madison County, Indiana*, p. 40.
17. Kellar, in Kellar (ed.), *Solon Robinson*, vol. 1, pp. 25–27; and Solon Robinson in ibid., pp. 265–72.
18. Solon Robinson, in Kellar (ed.), *Solon Robinson*, vol. 1, pp. 118–20.
19. Solon Robinson, in Kellar (ed.), *Solon Robinson*, vol. 1, p. 346 (also see pp. 348, 354–55); John Ade, *Newton County*, pp. 126–28; and Gates, *The Farmer's Age*, p. 180.
20. Power, *Planting Corn Belt Culture*, p. 55. The *Prairie Farmer* is an influential newspaper that was founded at Chicago in 1841 by Solon Robinson's friend John S. Wright. Its history is summarized in Schapsmeier and Schapsmeier, *Encyclopedia of American Agricultural History*, pp. 277–78.
21. Whitaker, *Feedlot Empire*, p. 21; and Gates, *Agriculture and the Civil War*, p. 181 table.
22. Gates, *Agriculture and the Civil War*, p. 180.
23. William Keating, quoted in Prince, *Wetlands of the American Midwest*, p. 119.
24. Prince, *Wetlands of the American Midwest*, pp. 131, 213–15. About early settlers hesitating to settle open prairies even if they were dry and fertile, see (among many accounts) Schramm, *The Schramm Letters*, p. 54; and Shannon, *The Farmer's Last Frontier*, p. 33.
25. John Ade, *Newton County*, p. 45 (also see pp. 41–45, 106–7); and George Ade, "Prairie Kings of Yesterday," p. 77. Such draining of shallow lakes and marshes toward the Kankakee River led to faster results once that river itself was straightened by dredging a channel through it, a project that started in the late 1800s. George Blakey tells the story in *Creating a Hoosier Self-Portrait*, pp. 159–64.

26. George Ade, "Prairie Kings of Yesterday," p. 76; and Prince, *Wetlands of the American Midwest*, p. 190.
27. Prince, *Wetlands of the American Midwest*, p. 212–13. By 1930, most of the counties in northern Indiana each held 10,000 to 100,000 acres of cropland underlain by tile drainage pipes (p. 232 map).
28. Killian, "Mint Farming in Lakeville," pp. 44–47 including diagram; and Phillips, *Indiana in Transition*, pp. 154–55.
29. Timothy Wright, quoted in Schob, *Hired Hands and Plowboys*, p. 25. A similar tribute comparing oxen favorably with horses appears in Oliver Johnson, *A Home in the Woods*, pp. 22–23. For facts alone see Gates, *The Farmer's Age*, pp. 227–28.
30. U.S. Census of 1870, *The Statistics of the Wealth and Industry of the United States*, pp. 82, 86–87, 90–91; and Ball, *Northwestern Indiana*, p. 127. Ball adds that where marshland still impinged on farmland "in Jasper and Newton and Starke, as newer counties and not feeling so soon the influence of the railroads, the use of oxen continued into later years."

Chapter Ten

1. Vincent, *Southern Seed, Northern Soil*, p. 88. Such inheritance that is divided more or less equally among all the children is called "partible" inheritance.
2. Rugh, *Our Common Country*, p. 114.
3. Rodgers, "Hoosier Women and the Civil War Home Front," p. 112. Another historian, however, asserts that in the northern U.S. overall, farm women *did* often step into men's farm roles. Gates, *Agriculture and the Civil War*, pp. 232, 242–43.
4. Interview of Ross and Avis Paulson, by author, pp. 1, 9, 11. The young John Cutler helped guard Lincoln at Gettysburg but he failed to hear any of Lincoln's speech. His unit formed a mounted cordon through which Lincoln's carriage reached the speakers' platform; the unit then swung its horses around and formed a square surrounding the platform, watching the crowd from horseback.
5. Rodgers, "Hoosier Women and the Civil War Home Front," p. 114; and pp. 105, 111–13.
6. Ibid., p. 105; and Madison, *The Indiana Way*, p. 326 table. Only the towns whose population exceeded 2,500 were counted as "urban."
7. Thornbrough, *Indiana in the Civil War Era*, pp. 390–91 tables; Houk, "A Century of Indiana Farm Prices," pp. 23, 50–51, 61–62; and U.S. Census of Agriculture, *Agriculture of the United States in 1860*, p. 43 table. Since corn, wheat, and other grains sell at widely varied prices at different times of the year, the prices here are yearly averages that Howard Jacob Houk calculated from monthly averages, which he had first calculated from weekly averages that he gathered from newspaper price quotations throughout Indiana, and which he "then combined into a state average by weighing each [of nine geographical] district price[s] in proportion to the acreage or production of that commodity in each district"

as compared to that commodity's acreage or production in each of the state's eight other districts. Houk, pp. 4–5.

8. Post, "The 'Agricultural Revolution' in the United States," p. 224. The 1853–1856 Crimean War had given some of that boost to U.S. wheat prices by stymieing Russian wheat exports from the Black Sea ports that served Russia's Ukrainian "breadbasket."
9. Gates, *Agriculture and the Civil War*, pp. 228–29. For state-by-state figures as of 1860 and 1870 for improved acreage, value of farms, value of farm machinery, and value of livestock, see Gates, p. 376 table.
10. Shannon, *The Farmer's Last Frontier*, pp. 140, 144; and Danbom, *Born in the Country*, pp. 111–12.
11. Gates, *Agriculture and the Civil War*, p. 233.
12. Post, "The 'Agricultural Revolution' in the United States," p. 220. Nonetheless, one-fourth of all the McCormick reapers sold between 1854 and 1859 were purchased jointly by two or more individuals, and apparently most of those purchasers planned to share their reaper's use. Atack and Bateman, *To Their Own Soil*, p. 200.
13. Gates, *Agriculture and the Civil War*, p. 234. Also see Hurt, *American Farm Tools*, pp. 44–47.
14. For the economic logic involved, see Gregson, "Rural Response to Increased Demand," p. 342; and Atack and Bateman, *To Their Own Soil*, pp. 194–200. For 1851 prices of farm implements, see Gates, *The Farmer's Age*, pp. 288–89. Prices as of selected years between 1860 and 1900 appear in Shannon, *The Farmer's Last Frontier*, p. 140 table.
15. Threshing crews and machines are described in detail in Rikoon, *Threshing in the Midwest*, chapters 2–5. Recall that the cost of threshers was far higher than the cost of reapers even before steam-powered threshers gained prominence in the 1880s.
16. Danhof, *Change in Agriculture*, p. 260. Regarding hay rakes, those contraptions didn't really work well until a design that reached the market around 1861, says Gates, *Agriculture and the Civil War*, p. 236.
17. Thornbrough, *Indiana in the Civil War Era*, pp. 379–81; Danhof, *Change in Agriculture*, pp. 274–75; and Shannon, *The Farmer's Last Frontier*, pp. 170–71.
18. Stoll, *Larding the Lean Earth*, p. 33. Also see pp. 19–41.
19. Gregson, "Rural Response to Increased Demand," pp. 342–44. Emphasis in original. Not just economic tradeoffs but other important tradeoffs are examined by Gates, *The Farmer's Age*, pp. 163–66.
20. Headlee, *The Political Economy of the Family Farm*, pp. 5, 34–35; and Post, "The 'Agricultural Revolution' in the United States," pp. 216–17.
21. Meyer, "Emergence of the American Manufacturing Belt," p. 154; Pudup, "From Farm to Factory," p. 204; and Gates, *Agriculture and the Civil War*, pp. 232–33,

239–41 including table. Gates adds that wartime crop prices prompted a large jump as well in the use of imported guano as fertilizer.

22. Pudup, "From Farm to Factory," p. 206.
23. Shannon, *The Farmer's Last Frontier*, p. 139.
24. Gates, *Agriculture and the Civil War*, pp. 238–39.
25. The proportion of the Midwest's staple farm products that were exported from the region jumped from just 27 percent in 1840 to well over 50 percent by 1860. Cayton and Onuf, *The Midwest and the Nation*, p. 39; and the U.S. Census of 1860, *Agriculture of the United States in 1860*, p. clxviii.
26. Rugh, *Our Common Country*, pp. 68–70.
27. Danhof, *Change in Agriculture*, p. 77. The average Indiana farm hand's wage in 1860 had been only $13.71 a month plus room and board. (Atack and Bateman, *To Their Own Soil*, pp. 241–43 including p. 242 table.) Besides pay, farm hands often expected a daily ration of whiskey. Etcheson, *The Emerging Midwest*, p. 86.
28. Where combines were in use by 1896, wheat's work time per acre had fallen way down to 3.3 hours (Shannon, *The Farmer's Last Frontier*, p. 143 table). But combines by then were not yet much used except in the West Coast states. In the Midwest, including Indiana, combines didn't grow common until the New Deal of the 1930s. What combines *combined* were reaping and threshing. Another way to get the grain was to harvest it with a "header" which left the stalk in the field, but header-harvested grain still had to be threshed, whereas combine-harvested grain was thereby already threshed. (Hurt, *American Farm Tools*, pp. 49–52, 77–83.) The reasons *why* Midwesterners adoped combines far later than West Coast and High Plains' farms are explained by Rikoon, *Threshing in the Midwest*, pp. 147–50.
29. Rugh, *Our Common Country*, p. 65. Evidence follows on pp. 65–70.
30. Faragher, *Sugar Creek*, p. 237, emphasis added.
31. Shannon, *The Farmer's Last Frontier*, p. 143 including table.
32. For Indiana's tax jump, Esarey, *A History of Indiana*, vol. 1, pp. 462–64. Regarding mortgages and their pervasiveness even before the Civil War, see Danhof, *Change in Agriculture*, pp. 78–87. On the ubiquity of mortgages following the Civil War, Shannon, *The Farmer's Last Frontier*, pp. 184–90.
33. Gates, *Landlords and Tenants on the Prairie Frontier*, p. 64.
34. Factors that influenced families' choices among these options are discussed in Rugh, *Our Common Country*, pp. 73–78.
35. See Headlee, *The Political Economy of the Family Farm*, pp. 19, 62–74, 82. Headlee's conclusion that families' risk-aversion tended to take precedence over their profit-maximization is based on the widespread purchase of reapers by families who harvested less than seventy-eight acres of wheat—seventy-eight being the average number of wheat acres above which it became likely to "pay" (in profit terms) to buy a reaper.

36. Prince, *Wetlands of the American Midwest*, pp. 156–57; and Danhof, *Change in Agriculture*, p. 106. Even before the Graduation Act was passed in 1854, Congress during the Mexican War had begun to pass veterans' benefits in the form of land warrants. Thereby, over 1.3 million acres of public land located in Indiana was given free to veterans between 1847 and 1853. Oberly, *Sixty Million Acres*, pp. 82–87 including tables; and Lebergott, "'O Pioneers'," pp. 43–44.
37. Prince, *Wetlands of the American Midwest*, pp. 143–48, 156–57; and Thornbrough, *Indiana in the Civil War Era*, pp. 365–66.
38. Stoll, *Larding the Lean Earth*, pp. 31–41, 189–94, 209–13.
39. Calculated from Prince, *Wetlands of the American Midwest*, p. 149 table. Indiana experienced population jumps of 133 percent in the 1820s, 100 percent in the 1830s, 44 percent in the 1840s, 37 percent in the 1850s, 24 percent in the 1860s, and 18 percent in the 1870s. Also see Prince's p. 193 table.
40. Thornbrough, *Indiana in the Civil War Era*, pp. 390–91 tables; Prince, *Wetlands of the American Midwest*, p. 192; and Phillips, *Indiana in Transition*, pp. 148–50 including tables. Granted, even that much wheat was a major response to commercial incentives, since back in 1849 Indiana had produced only 6.3 bushels of wheat per capita—as compared that year to 53.6 bushels of corn per capita. Indeed, ten years earlier (in 1839) Indiana's wheat harvest had been far exceeded by its oat harvest. Carmony, *Indiana, 1816–1850*, p. 51.
41. Pudup, "From Farm to Factory," p. 205; and Shannon, *The Farmer's Last Frontier*, pp. 132–33, 136–37, 143 table. Descriptions of changing corn implements appear in Jones, *History of Agriculture in Ohio*, pp. 272–74. Analysis as well as illustrations can be found in Hurt, *American Farm Tools*, chapter 6; and also in Cochran, *The Development of American Agriculture*, pp. 191–96. The dating of silos' popularity is from Phillips, *Indiana in Transition*, pp. 150–51, and from Shannon, p. 137.
42. Madison, *The Indiana Way*, p. 147. These percentages include not just farm owner-operators but farm tenant-operators (including sharecroppers) and also full-time farm hands. For details about harvesting with corn sleds, corn binders, and corn shockers, all three of which achieved efficiency in the late 1800s, again see Hurt, *American Farm Tools*, chapter 6 (including illustrations).
43. Madison, *The Indiana Way*, p. 263.
44. Douglas Harper, *Changing Works*, pp. 137–57 provides a detailed analysis of "the corn revolution" and its effects. Another aspect of that "revolution" was an enormous expansion of corn production per acre through the use of hybrid seeds, which went mainstream in the 1930s when the New Deal restricted the number of acres that farmers could allot to corn.

Chapter Eleven

1. Hosea Smith Letters, 1810–1814, in the Indiana Historical Society archives, quoted in Rohrbough, *The Trans-Appalachian Frontier*, p. 94. As of 1810, Knox

County was still about ten times larger than it is today. For a series of maps showing Indiana's country boundary lines as of each dicennial census in the 1800s, see Thorndale and Dollarhide, *Map Guide to the U.S. Federal Censuses, 1790–1920*, pp. 106–13.

2. Caleb Lownes, 1815, in McCord (compiler), *Travel Accounts of Indiana*, p. 72.
3. William Faux, quoted in Thompson and Madigan, *One Hundred and Fifty Years of Indiana Agriculture*, p. 8.
4. Rohrbaugh, *The Trans-Appalachian Frontier*, p. 169. This was a boy in one of the three families mentioned in chapter 7 on pages 73–74, who left the Cincinnati area together due to depressed conditions there after the financial crash of 1819.
5. Wilkinson (ed.), "'To do for my self,'" pp. 401–3 including map. The sequence of settlement northward up the Wabash River is narrated by Esarey, *A History of Indiana*, vol. 1, pp. 273–78.
6. 13 August 1830, in the William McCutcheon Letters at the Indiana State Library, quoted in Rohrbough, *The Trans-Appalachian Frontier*, p. 172.
7. Kirby, "Rural Culture in the American Middle West," p. 586. Also see Duncan, "Old Settlers," p. 396.
8. Gates, *The Farmer's Age*, pp. 73–74, 81. Ironies in the emergence of sharper class distinctions are mulled over by Cayton and Onuf, *The Midwest and the Nation*, pp. 30–34. Sharper class lines had also emerged in Kentucky under similar circumstances: "When there remained no more unclaimed farmland to take up, . . . two classes of people emerged." (Ardrey, *Welcome the Traveler Home*, pp. 18–19.) Also regarding the emergence of class lines in early Kentucky, see Perkins, *Border Life*, pp. 121–23.
9. Swiss immigrant Oswald Ragatz [1842], quoted in Rohrbough, "Diversity and Unity in the Old Northwest," p. 81.
10. Rohrbough, "Diversity and Unity in the Old Northwest," p. 78. Also see Vincent, *Southern Seed, Northern Soil*, pp. 56–60; Danhof, *Change in Agriculture*, pp. 74–78; and Schob, *Hired Hands and Plowboys*, pp. 5–20 and passim.
11. Carter, "Rural Indiana in Transition," p. 114. The ex-congressman did agree to supply the bricks for the chimney.
12. Prince, *Wetlands of the American Midwest*, p. 193. As mentioned earlier, persons of foreign birth accounted in 1860 for 19 percent of Indiana's "heads of household." (Atack and Bateman, *To Their Own Soil*, p. 74 table.) For examples of foreign-born farm hands, see Taylor and McBirley (eds.), *Peopling Indiana*, especially pp. 479–84 in the chapter by Winquist on "Scandinavians."
13. Schob, *Hired Hands and Plowboys*, pp. 173–202.
14. Schob, *Hired Hands and Plowboys*, p. 188.
15. John G. Clark, *The Grain Trade in the Old Northwest*, p. 150. For a map of Indiana's major railroads as of 1860, see Cochran, *The Development of American Agriculture*, p. 219. A canal map is also there at p. 215. Indiana's railroads twenty

years later, in 1880, are mapped and identified in Madison, *The Indiana Way*, p. 156.

16. Gates, *The Farmer's Age*, pp. 165–69.
17. Vincent, *Southern Seed, Northern Soil*, p. 86.
18. Calculated from the U.S. Census of 1850, *Statistical View of the United States*, p. 169; U.S. Census of 1860, *Agriculture of the United States in 1860*, p. 198; and U.S. Census of 1870, *The Statistics of the Wealth and Industry of the United States*, pp. 81, 86, 90, 350. For descriptions of the major new farm implements, see Thornbrough, *Indiana in the Civil War Era*, pp. 373–79.
19. Vincent, *Southern Seed, Northern Soil*, pp. 41, 84. Note that in the year 1832 the government had reduced the minimum size of parcel that it would sell from 80 acres to 40 acres.
20. Calculated from the U.S. Census of 1870, *The Statistics of Wealth and Industry of the United States*, pp. 81, 86, 90. These figures include improvements on farms such as buildings. Also see Vincent, *Southern Seed, Northern Soil*, p. 84. Vincent discusses the consequences of rising land and equipment costs on pp. 112–15.
21. Ezra C. Seaman, 1868, quoted in Danhof, *Change in Agriculture*, p. 5 n. 10. During just the 1850s, the total cash value of Indiana farms rose from $136 million in 1850 to $356 million in 1860. U.S. Census of 1870, *The Statistics of the Wealth and Industry of the United States*, pp. 86, 90.
22. Vincent, *Southern Seed, Northern Soil*, pp. 57–61, 86–87; and Rohrbough, "Diversity and Unity in the Old Northwest," pp. 83–84.
23. James M. Allen [of Beechymire, Union County, Indiana] to J.D. Davidson, 20 January 1856, quoted in Carter, "Rural Indiana in Transition," pp. 110–11.
24. Cayton and Onuf, *The Midwest and the Nation*, pp. 34–39.
25. Taylor, *The Transportation Revolution*, p. 79 table; Thornbrough, *Indiana in the Civil War Era*, p. 361 table; Carter, "Rural Indiana in Transition," p. 113; U.S. Census of 1850, *Statistical View of the United States*, p. 169; and U.S. Census of 1860, *Agriculture of the United States in 1860*, p. 42. "Improved land" included crop fields, pasture fields, orchards, and a few other miscellaneous categories of use which entailed land-preparation. During the 1860s, Indiana's improved farm acres continued climbing, reaching 10,104,279 acres in 1870, which was almost half the land surface of the state. U.S. Census of 1870, *The Statistics of the Wealth and Industry of the United States*, p. 81. But by that time (1870) the motive for clearing new land was often simply to persevere rather than to profit.
26. U.S. Census of 1860, *Preliminary Report on the Eighth Census, 1860* [1862], p. 196 table.
27. Shannon, *The Farmer's Last Frontier*, pp. 295–302.
28. Berry, *Western Prices before* 1861, p. 226; and Earle, *Geographical Inquiry and American Historical Problems*, p. 312. And by that national crisis year of 1860 the quantity of grain and flour shipped from Chicago had grown to be enormous.

The first omen had been the leap from 40,000 bushels in 1841 to 586,907 bushels in 1842. Another jump went from 6,422,181 bushels in 1853 to 12,902,320 bushels in 1854. And a third leap would top 1860's 31,109,059 bushels with 50,511,862 bushels in 1861. U.S. Census of 1860, *Agriculture of the United States in 1860*, p. cxlix.

29. See Rockoff, *The Free Banking Era*; and Thornbrough, *Indiana in the Civil War Era*, pp. 428–32. The right to free, unrestricted entry into the banking business was mandated by Indiana's 1851 constitution—conditional of course on the sale of enough bank stock to show that investors trusted a bank to fulfill its obligations. (In practice, Indiana banks often began doing business before their paid-in capital reached the amount required by their charter from the state. But subsequent scrutiny has shown the banks did basically limit the issuing of their own currency to the ratio that was set by law between their currency in circulation and their paid-in bank stock. Adams, "The Role of Banks in the Economic Development of the Old Northwest," p. 225.)
30. Phillips, *Indiana in Transition*, pp. 30–31. For both farm sizes and production figures, see the U.S. Census of 1850, *Statistical View of the United States*, p. 169; U.S. Census of 1860, *Agriculture of the United States in 1860*, p. 198; and U.S. Census of 1870, *The Statistics of the Wealth and Industry of the United States*, pp. 81–91, 350. For the specifics of falling prices for farm products by 1870, falls which in following years were often drastic, see Houk, "A Century of Indiana Farm Prices," pp. 42–66 tables.
31. Thornbrough, *Indiana in the Civil War Era*, pp. 389–90. The specifics of rising prices for farm products in the 1850s appear in Houk, "A Century of Indiana Farm Prices," pp. 42–66 tables.
32. The fourth major reason was the Crimean War of 1853–56, which inhibited grain shipments from Russia's Ukrainian "breadbasket" to western Europe. See Prince, *Wetlands of the American Midwest*, p. 191; and Gates, *The Farmer's Age*, pp. 166–67.
33. Carter, "Rural Indiana in Transition," p. 108; and U.S. Census of 1860, *Agriculture of the United States in 1860*, p. clxiv.
34. Earle, "Regional Economic Development West of the Appalachians," p. 181. But the 1850s then witnessed a rapid reorientation toward the East, and by 1860 very few of the Midwest's exports and imports were traveling via New Orleans (p. 184).
35. U.S. Census of 1860, *Agriculture of the United States in 1860*, p. clxvi. Details of this three-way competition for farmers' transportation business, along with the transportation price-cutting that it fostered, appear in Benton, *The Wabash Trade Route*, pp. 101–8.
36. Esarey, *A History of Indiana*, vol. 2, p. 861.
37. U.S. Census of 1860, *Agriculture of the United States in 1860*, p. clxvi. On railroad consolidation in Indiana, see Thornbrough, *Indiana in the Civil War Era*, pp.

346–52; and Phillips, *Indiana in Transition*, pp. 228–47 including maps. Unsuccessful efforts in Indiana's legislature to limit the power of railroad monopolies are described by Thornbrough, pp. 352–61.

38. Meyer, "Emergence of the American Manufacturing Belt," p. 151. Average ton-mile freight rates in the U.S. from 1780 to 1900 are graphed in Cochrane, *The Development of American Agriculture*, p. 216.
39. In actual practice, Upland Southerners and their midwestern descendants generally transferred their land to their children *before* they died, but the principle of rough equality reigned in that case too. Nicholson, "Swine, Timber, and Tourism," pp. 74–75; and Rugh, *Our Common Country*, pp. 73–78.
40. Appleby, *Capitalism and a New Social Order*, pp. 99–100. Also see pp. 39–50, 88–105 there. And see Andrew Shankman, *Crucible of American Democracy*, pp. 3–9, 12. But Shankman additionally highlights some ambiguities in Jeffersonian economic thought which later led to the Democrat-Whig bifurcation in the 1830s (pp. 233–46).
41. See Kirby, "Rural Culture in the American Middle West," pp. 587–88; and Shankman, *Crucible of American Democracy*, pp. 220–24, 227–33.
42. And yet those two exploited groups couldn't agree with each other—because self-employed farmers wanted more money put into circulation, which would help them pay off their debts, whereas wage workers *didn't* want more money put into circulation since that would lower the buying power of their wages, and they didn't relish the trauma of going on strike just to keep up with inflation.
43. Vincent, *Southern Seed, Northern Soil*, pp. 84–88. For details about the proliferation of farm machinery, see Thornbrough, *Indiana in the Civil War Era*, pp. 373–79.
44. Houk, "A Century of Indiana Prices," pp. 50–51 table. Other price reports vary slightly. See for instance Vincent, *Southern Seed, Northern Soil*, pp. 81, 86, 113; Thornbrough, *Indiana in the Civil War Era*, pp. 389–90 (including table), 393; and Thompson and Madigan, *One Hundred and Fifty Years of Indiana Agriculture*, p. 20. At Cincinnati meanwhile, the prices paid for a bushel of corn were 12 cents in 1826, 32 cents in 1835, 37 cents in 1853, and 48 cents in 1860. U.S. Census of 1860, *Agriculture of the United States in 1860*, p. clxviii. Yet the relative *viability* of the corn belt is shown by the fact that, straight through from the 1870s to the 1930s, corn rose steadily in value relative to wheat, overall almost doubling in its value relative to wheat during those sixty years. (Houk, "A Century of Indiana Farm Prices," p. 25 table and graph.) Corn's double role as both human and animal food helps explain that.
45. Esarey, *A History of Indiana*, vol. 2, pp. 859–62; Thornbrough, *Indiana in the Civil War Era*, pp. 285–317, 352–61, 397–401; and Phillips, *Indiana in Transition*, pp. 30–40.
46. Gates, *The Farmer's Age*, p. 71 graph.

47. Martin, "The Economic Transformation of Indiana Agriculture," p. 34 graph.
48. Blanke, *Sowing the American Dream*, p. 18; and Vedder and Gallaway, "Migration and the Old Northwest," p. 163 table. Blanke adds that by 1870, Illinois too was losing more out-migrants than it was gaining in-migrants. In both states, of course, the population kept rising due to natural increase.
49. Thornbrough, *Indiana in the Civil War Era*, p. 541.
50. Shaw, "The Progress of the World" (October 1935), p. 17.
51. Higgs, *The Transformation of the American Economy*, p. 83 map. Also see Prince, *Wetlands of the American Midwest*, pp. 191–92; Shannon, *The Farmer's Last Frontier*, p. 163 table; and Phillips, *Indiana in Transition*, pp. 148–50 including tables.
52. Williams, *The Roots of the Modern American Empire*, pp. 158–71, 318–26.
53. Buley, *The Old Northwest*, vol. 1, p. 194. Buley was here summarizing an 1833 article in a farm journal published at Cincinnati. Also see Buley, vol. 1, pp. 168–69, 193–94.
54. Quoted in Thompson and Madigan, *One Hundred and Fifty Years of Indiana Agriculture*, p. 17. Also see Thornbrough, *Indiana in the Civil War Era*, pp. 379–81, 395–97; and Carmony, *Indiana, 1816–1850*, pp. 63–66.
55. Quoted in Thornbrough, *Indiana in the Civil War Era*, p. 379. For equally early warnings in Illinois, see Shannon, *The Farmer's Last Frontier*, pp. 169–70.
56. For the high corn and pork prices in the 1850s and '60s, followed by their fall starting around 1870, see Houk, "A Century of Indiana Farm Prices," pp. 50–51, 55 tables. A disheartening account of agricultural adversity and subsequent social tensions in the 1870s and '80s in south-central Indiana appears in Nicholson, "Swine, Timber, and Tourism," pp. 134–42.
57. Calculated from Thornbrough, *Indiana in the Civil War Era*, p. 393 table. Corn prices from Houk, "A Century of Indiana Farm Prices," pp. 50–51 table.

Epilogue

1. Parker, *Europe, America, and the Wider World*, vol. 2, p. 278.
2. Rose, "Upland Southerners," pp. 259–62; Rose, "Hoosier Origins," p. 227 map; and Bigham, *Towns and Villages of the Lower Ohio*, pp. 26–41.
3. Rose, "Upland Southerners," p. 261.
4. One benefit due largely to USDA initiatives is that the annual rate of soil erosion from Indiana cropland continues to decline, and as of 1997 had fallen to an estimated 3.5 tons per year per average acre under active cultivation. U.S. Dept. of Agriculture, Natural Resources Conservation Service, Summary Report, 1997 National Resources Inventory, Revised December 2000, Tables 10 and 11—accessed at www.nrsc.usda.gov/technical/NRI/1997/summary_report/table 10.html (and table 11)
5. Neth, *Preserving the Family Farm*, p. 13. Also see Danbom, *Born in the Country*, pp. 248–49.

6. Regarding the broad context of this social loss, see Carnes, *Secret Ritual and Manhood in Victorian America*, pp. 106–13. William N. Parker thinks that, even among *farm* families, the family itself lost its role as "the principal locus of the transmittal of farming techniques to the succeeding generation [because] schools, colleges, and extension services [4-H clubs] usurped its function" (Parker, *Europe, America, and the Wider World*, vol. 2, p. 178.). But Parker surely overestimates such non-family influences.
7. Cayton and Onuf, *The Midwest and the Nation*, p. xvii.
8. Among other sources, Sauer, "Homestead and Community on the Middle Border," p. 37.
9. Jefferson [1805], quoted in Appleby, *Capitalism and a New Social Order*, p. 99.
10. Danbom, *Born in the Country*, p. 251. Among many first-hand accounts of traditional childhoods grounded in farm chores, see Jager, *The Fate of Family Farming*, pp. 24–28. Jager's memoir *Eighty Acres* is devoted entirely to that subject. He grew up between the mid–1930s and early 1950s in a farm neighborhood of second-generation Dutch families in southwestern Michigan.
11. Among many accounts, see James, *Money and Captial Markets in Postbellum America*, especially pp. 12–13, 17 (graph), 74–78, 107–10 (including graphs), 189–93.
12. One aim of the Grange was to bypass business monopolies through farmers' cooperatives. But how that attempt faltered and fell in the 1870s for the Grange chapters in east-central Indiana's Delaware County is analyzed by Blanke, *Sowing the American Dream*, pp. 117–23. Also see pp. 221–25. For statewide Grange data, see Thornbrough, *Indiana in the Civil War Era*, pp. 401–3; Thompson and Madigan, *One Hundred and Fifty Years of Indiana Agriculture*, pp. 23–24; and Esarey, *A History of Indiana*, vol. 2, pp. 851–55.
13. For "parity" details see Schnapsmeier and Schnapsmeier, *Encyclopedia of American Agricultural History*, pp. 261–62. During World War I itself, the purchasing power that was conferred on Indiana farm families by the prices they received for farm products rose even higher. See *Prices of Indiana Farm Products, 1841–1955*, p. 9 graph.
14. Danbom, *Born in the Country*, pp. 161–64; and Parker, *Europe, America, and the Wider World*, vol. 2, pp. 172–76. Like the late 1800s' dark clouds, those of the 1920s were again spawned by agricultural overproduction. And again in the 1920s (as in the late 1800s) the plowing of new land was a major factor in causing overproduction, but an additional reason in the 1920s was the use of tractors. Parker points out that "the elimination of hay and oats consumed by farmwork animals added ninety million acres [nationwide] to the land available for food crops and accounts for over half the increase in net farm output between 1920 and 1940." Parker, vol. 2, p. 222 n. 12. Also see p. 173. (Net farm output means farm output that left the farm.)

15. Danhof, *Change in Agriculture*, p. 3; and U.S. Census Bureau, *Historical Statistics of the United States*, Part 1, pp. 8, 457.
16. Quoted in "A Lot Rides on WTO Discussions," Gannett News Service, 25 November 1999.

Works Cited

UNPUBLISHED SOURCES

The Ewing Brothers Papers. [W.G. and G.W. Ewing.] Manuscript Section of the Indiana Division, Indiana State Library, Indianapolis. Index No. B-119.

Indiana Federal Writers' Project/Program Papers. Indiana State University Library, Special Collections. [See also the guide to this archive, published under the same title; compiled by Robert L. Carter, edited by David E. Vancil. Terre Haute: Friends of the Cunningham Memorial Library, 1992.]

Papers of Indiana Governor James P. Goodrich. Herbert Hoover Presidential Library, West Branch, Iowa.

Interview of Ross and Avis Paulson by the author. Conducted on 26 June 1996 at Rock Island, Illinois. Transcript in author's possession.

The David W. Walton Blacksmith Shop Account Book, 1830–1842 [1830–35 in Tippecanoe County, Indiana; 1835–42 in Cedar County, Iowa]. Microfilm in University of Iowa Library, Special Collections. Original in possession of J. Curtis Frymour, Wilton Junction, Iowa.

The Paul Weer Papers. Indiana Historical Society Library, Indianapolis. M 0293. Index No. A-393.

PUBLISHED WORKS

Abernethy, Thomas Perkins. *From Frontier to Plantation in Tennessee: A Study in Frontier Democracy*. Chapel Hill: University of North Carolina Press, 1932.

———. *Three Virginia Frontiers*. Baton Rouge: Louisiana State University Press, 1940; reprinted Gloucester, MA: Peter Smith, 1962.

Adams, Donald R., Jr. "The Role of Banks in the Economic Development of the Old

Northwest." In *Essays in Nineteenth Century Economic History: The Old Northwest*, ed. David C. Klingaman and Richard K. Vedder. Athens: Ohio University Press, 1975.

Ade, George. "Prairie Kings of Yesterday." *Saturday Evening Post*, 4 July 1931, pp. 14, 75–78.

Ade, John. *Newton County [1853–1911]*. Indianapolis: Bobbs-Merrill, 1911.

Aley, Ginette. "Grist, Grit, and Rural Society in the Early Nineteenth Century Midwest." *Ohio Valley History* 5 (Summer 2005): 3–20.

Anderson, Oscar Edward, Jr. *Refrigeration in America: A History of a New Technology and Its Impact*. Princeton: Princeton University Press, 1953.

Anderson, Terry L. "The First Privatization Movement." In *Essays on the Economy of the Old Northwest*, ed. David C Klingaman and Richard K. Vedder. Athens: Ohio University Press, 1987.

Androit, Jay (compiler). *Township Atlas of the United States*. McLean, VA: Documents Index, 1991.

Ankli, Robert E. "Farm-Making Costs in the 1850s." In *Farming in the Midwest, 1840–900*, ed. James W. Whitaker. Washington, DC: Agricultural History Society, 1974.

Appleby, Joyce. *Capitalism and a New Social Order: The Republican Vision of the 1790s*. New York: New York University Press, 1984.

Ardrey, Julia S. (ed.). *Welcome the Traveler Home: Jim Garland's Story of the Kentucky Mountains*. Lexington: University Press of Kentucky, 1983.

Argersinger, Peter H., and JoAnn E. Argersinger. "The Machine Breakers: Farmworkers and Social Change in the Rural Midwest of the 1870s." *Agricultural History* 58 (1984): 393–410.

Arndt, Karl J. R. (ed.). *A Documentary History of the Indiana Decade of the Harmony Society, 1814–1824*. 2 vols. Indianapolis: Indiana Historical Society, 1975.

Asch, David L., and Nancy B. Asch. "Prehistoric Plant Cultivation in West-Central Illinois." In *Prehistoric Food Production in North America*, ed. Richard I. Ford. Anthropology Paper No. 75. Ann Arbor: Museum of Anthropology, University of Michigan, 1985.

Atack, Jeremy, and Fred Bateman. "Self-Sufficiency and the Marketable Surplus in the Rural North, 1860." *Agricultural History* 58 (1984): 296–313.

———. *To Their Own Soil: Agriculture in the Antebellum North*. Ames: Iowa State University Press, 1987.

———. "Yankee Farming and Settlement in the Old Northwest: A Comparative Analysis." In *Essays on the Economy of the Old Northwest*, ed. David C. Klingaman and Richard K. Vedder. Athens: Ohio University Press, 1987.

Baer, M. Teresa, Kathleen M. Breen, and Judith Q. McMullen (eds.). *Centennial Farms of Indiana*. Indianapolis: Indiana Historical Society, 2003.

Baker, Raymond. "Indian Corn and Its Culture." In *Farming in the Midwest, 1840–1900*, ed. James W. Whitaker. Washington, DC: Agricultural History Society, 1974.

Ball, T. H. *Northwestern Indiana from 1800 to 1900*. Chicago: Donohue & Henneberry, 1900.

Banner, Stuart. *How the Indians Lost Their Land: Law and Power on the Frontier*. Cambridge: Belknap Press of Harvard University Press, 2005.

Barce, Elmore. *The Land of the Miamis*. Fowler, IN: The Benton Review Shop, 1922.

Barnhart, John D. *Valley of Democracy: The Frontier versus the Plantation in the Ohio Valley, 1775–1818*. Bloomington: Indiana University Press, 1953.

Barnhart, John D., and Donald F. Carmony. *Indiana: From Frontier to Industrial Commonwealth*. 2 vols. New York: Lewis Publishing Company, 1954.

Barnhart, John D., and Dorothy L. Riker. *Indiana to 1816: The Colonial Period*. Indianapolis: Indiana Historical Bureau and Indiana Historical Society, 1971.

Beckwith, Hiram W. *The Illinois and Indiana Indians*. Chicago: Fergus Printing Company, 1884.

Benton, Elbert Jay. *The Wabash Trade Route in the Development of the Old Northwest*. Baltimore: The Lord Baltimore Press, 1903.

Berkhofer, Robert F., Jr. "Cultural Pluralism versus Ethnocentrism in the New Indian History." In *The American Indian and the Problem of History*, ed. Calvin Martin. New York: Oxford University Press, 1987.

Berry, Thomas Senior. *Western Prices before 1861: A Study of the Cincinnati Market*. Cambridge: Harvard University Press, 1943.

Berry, Wendell. *The Unsettling of America: Culture and Agriculture*. New York: Avon, 1978.

Berthrong, Donald J. *Indians of Northern Indiana and Southwestern Michigan*. New York: Garland, 1974.

Bigham, Darrel E. *Towns and Villages of the Lower Ohio*. Lexington: University Press of Kentucky, 1998.

Billington, Ray Allen. *Westward Expansion: A History of the American Frontier*. New York: Macmillan, 1974.

Birkbeck, Morris. *Notes on a Journey to America*. 1818; reprinted Ann Arbor: University Microfilms, Inc., 1966.

Blakey, George T. *Creating a Hoosier Self-Portrait: The Federal Writers' Project in Indiana, 1935–1942*. Bloomington: Indiana University Press, 2005.

Blanke, David. *Sowing the American Dream: How Consumer Culture Took Root in the American Midwest*. Athens: Ohio University Press, 2000.

Boas, Franz. *Primitive Art*. New York: Dover Publications, 1955.

Boserup, Ester. *The Conditions of Agricultural Growth*. Chicago: Aldine, 1965.

Brelsford, Bridgie Brill. *Indians of Montgomery County, Indiana*. Crawfordsville, IN: Montgomery County Historical Society, 1985.

Briggs, Winstanley. "Le Pays des Illinois." *William and Mary Quarterly*, 3rd series, 47 (1990): 130–156.

Bromme, Traugott. "The State of Indiana" (1848). Translated by Richard L. Bland. *The Hoosier Genealogist* 44, no. 3 (Fall 2004): 138–143.

Brown, James A. "The Impact of the European Presence on Indian Culture." In *Contest for Empire, 1500–1775*, ed. John B. Elliott. Indianapolis: Indiana Historical Society, 1975.

Brown, Samuel R. *The Western Gazetteer; or Emigrant's Directory* [1817], pages 38–80 thereof only. In *Indiana as Seen by Early Travelers*, ed. Harlow Lindley. Indianapolis: Indiana Historical Commission, 1916.

Buley, R. Carlyle. *The Old Northwest Pioneer Period, 1815–1840*, 2 vols. Indianapolis: Indiana Historical Society, 1950.

Burnet, Jacob. *Notes on the Early Settlement of the North-Western Territory*. New York: D. Appleton & Co., 1847; reprinted New York: Arno Press, 1975.

Bushman, Richard L. "Family Security in the Transition from Farm to City, 1750–1850." *Journal of Family History* 6 (1981): 238–256.

Butterworth, C.W. (ed.). *The Washington-Crawford Letters*. Cincinnati: Robert Clarke & Co., 1877.

Cameron, Rondo. *A Concise Economic History of the World: From Paleolithic Times to the Present*. New York: Oxford University Press, 1989.

Carmony, Donald F. *Indiana, 1816–1850: The Pioneer Era*. Indianapolis: Indiana Historical Bureau and Indiana Historical Society, 1998.

———, ed., "From Lycoming County, Pennsylvania, to Parke County, Indiana: Recollections of Andrew Ten Brook, 1786-1823." *Indiana Magazine of History* 61 (1965): 1–30.

Carnes, Mark C. *Secret Ritual and Manhood in Victorian America*. New Haven: Yale University Press, 1989.

Carter, Harvey L. "Rural Indiana in Transition, 1850-1860." *Agricultural History* 20 (1946): 107–121.

Cass, Lewis. *Considerations on the Present State of the Indians, and Their Removal to the West of the Mississippi*. Boston: Gray and Bowen, 1828; reprint New York: Arno Press, 1975.

Cayton, Andrew R. L. *Frontier Indiana*. Bloomington: Indiana University Press, 1996.

———. "The Northwest Ordinance from the Perspective of the Frontier." In *The Northwest Ordinance, 1787: A Bicentennial Handbook*, ed. Robert M. Taylor, Jr. Indianapolis: Indiana Historical Society, 1987.

Cayton, Andrew R. L., and Peter S. Onuf. *The Midwest and the Nation: Rethinking the History of an American Region*. Bloomington: Indiana University Press, 1990.

Chandler, Alfred N. *Land Title Origins: A Tale of Force and Fraud*. New York: Robert Schalkenbach Foundation, 1945.

Christian, David. *Maps of Time: An Introduction to Big History*. Berkeley: University of California Press, 2004.

Clark, John G. *The Grain Trade in the Old Northwest*. Urbana: University of Illinois Press, 1966.

Clark, Thomas D. "The Advance of the Anglo-American Frontier, 1700–1783." In

Contest for Empire, 1500–1775, ed. John B. Elliott. Indianapolis: Indiana Historical Society, 1975.

Claster, Jill. *The Medieval Experience, 1300–1400*. New York: New York University Press, 1982.

Clifton, James A. *The Prairie People: Continuity and Change in Potawatomi Indian Culture, 1665–1965*. Lawrence: The Regents Press of Kansas, 1977.

Cochran, Willard W. *The Development of American Agriculture: A Historical Analysis*. Minneapolis: University of Minnesota Press, 1979.

"Colonus." In *Dictionary of the Middle Ages*, 13 vols. Vol. 3. New York: Charles Scribner's Sons, 1983.

Cord, Xenia E. "Black Rural Settlements in Indiana before 1860." In *Indiana's African-American Heritage: Essays from* Black History News and Notes, ed. Wilma L. Gibbs. Indianapolis: Indiana Historical Society, 1993.

Cowan, C. Wesley. "Understanding the Evolution of Plant Husbandry in Eastern North America: Lessons from Botany, Ethnography and Archaeology." In *Prehistoric FoodProduction in North America*, ed. Richard I. Ford. Ann Arbor: Museum of Anthropology, University of Michigan, Anthropology Paper No. 75, 1985.

Cox, Stephen. "Back to the 1820s: The Re-Restoration." *Traces of Indiana and Midwestern History* 5, no. 1 (Winter 1993): 28–33.

———. "New Life: Eli Lilly and the First Restoration." *Traces of Indiana and Midwestern History* 5, no. 1 (Winter 1993): 24–27.

Cronon, William. *Nature's Metropolis: Chicago and the Great West*. New York: W.W. Norton, 1991.

Crumrin, Timothy. "Between Two Worlds: William Conner of Indiana." *Traces of Indiana and Midwestern History* 5, no. 1 (Winter 1993): 18–23.

Danhof, Clarence H. *Change in Agriculture: The Northern United States, 1820–1870*. Cambridge: Harvard University Press, 1969.

Danbom, David B. *Born in the Country: A History of Rural America*. Baltimore: Johns Hopkins University Press, 1995.

David, Paul A. "The Mechanization of Reaping in the Antebellum Midwest." In *Technical Change, Innovation and Economic Growth*, ed. Paul A. David. Cambridge, Eng.: Cambridge University Press, 1975.

Dean, Thomas. *Journal of Thomas Dean: A Voyage to Indiana in 1817*. Indianapolis: John Candee Dean, 1918.

Dillon, John B. "The National Decline of the Miami Indians." In *Proceedings of the Indiana Historical Society, 1830–1886*. Indiana Historical Society Publications, Vol. 1, No. 1. Indianapolis: The Bowen-Merrill Company, 1897.

Dillon, Lowell I., and Edward E. Lyon. *Indiana: Crossroads of America*. Dubuque, IA: Kendall/Hunt Publishing Co., 1978.

Doolittle, William E. *Cultivated Landscapes of Native North America*. Oxford: Oxford University Press, 2006.

Dufour, Perret. *The Swiss Settlement of Switzerland County Indiana*. Indianapolis: Indiana Historical Commission, 1925.

Duncan, Robert B. "Old Settlers." In *Indiana Historical Society Publications*. Vol. 2, no. 10. Indianapolis: Bowen-Merrill Company, 1894.

Earle, Carville. *Geographical Inquiry and American Historical Problems*. Stanford: Stanford University Press, 1992.

———. "Regional Economic Development West of the Appalachians, 1815–1860." In *North America: The Historical Geography of a Changing Continent*, ed. Robert D. Mitchell and Paul A. Groves. Towowa, NJ: Rowman and Littlefield, 1987.

Easterlin, Richard A. "Farm Production and Income in Old and New Areas at Mid-Century." In *Essays in Nineteenth Century Economic History: The Old Northwest*, ed. David C. Klingaman and Richard K. Vedder. Athens: Ohio University Press, 1975.

Ebeling, Walter. *The Fruited Plain: The Story of American Agriculture*. Berkeley: University of California Press, 1979.

Edmunds, R. David. *The Potawatomis: Keepers of the Fire*. Norman: University of Oklahoma Press, 1978.

Eisenhower, Dwight D. *At Ease: Stories I Tell to Friends*. New York: Doubleday, 1967.

Ekberg, Carl J. *French Roots in the Illinois Country: The Mississippi Frontier in Colonial Times*. Urbana: University of Illinois Press, 1998.

Ellsworth, Henry William. *Valley of the Upper Wabash, Indiana, with Hints on Its Agricultural Advantages*. New York: Pratt, Robinson, and Co., 1838; reprinted New York: Arno Press, 1975.

Elsmere, Jane Shaffer. *Henry Ward Beecher: The Indiana Years*. Indianapolis: Indiana Historical Society, 1973.

Encyclopedia of Appalachia. Edited by Rudy Abramson and Jean Haskell. Knoxville: University of Tennessee Press, 2006.

Encyclopedia of the United States in the Nineteenth Century, 3 vols. New York: Charles Scribner's Sons, 2001.

Esarey, Logan. *A History of Indiana*. 2 vols. in one. Indianapolis: Hoosier Heritage Press, 1970. [Note that this edition's page numbers differ from those of earlier editions.]

———. *The Indiana Home*. Bloomington: Indiana University Press, 1953.

Escobar, Arturo. *Encountering Development: The Making and Unmaking of the Third World*. Princeton: Princeton University Press, 1995.

Etcheson, Nicole. *The Emerging Midwest: Upland Southerners and the Political Culture of the Old Northwest, 1787–1861*. Bloomington: Indiana University Press, 1996.

Faragher, John Mack. "Americans, Mexicans, Métis: A Community Approach to the Comparative Study of North American Frontiers." In *Under an Open Sky: Rethinking America's Western Past*, ed. William Cronon, et al. New York: Norton, 1992.

———. "The Midwestern Farming Family, 1850." In *Women's America: Refocusing the Past.* Fourth edition. Ed. Linda K. Kerber and Jane Sharon De Hart. New York: Oxford University Press, 1995.

———. "Open-Country Community: Sugar Creek, Illinois, 1820–1850." In *The Countryside in the Age of Capitalist Transformation*, ed. Steven Hahn and Jonathan Prude. Chapel Hill: University of North Carolina Press, 1985.

———. *Sugar Creek: Life on the Illinois Prairie.* New Haven: Yale University Press, 1986.

Fatout, Paul. *Indiana Canals.* West Lafayette: Purdue University Studies, 1972.

Faulkner, Charles H. *The Late Prehistoric Occupation of Northwestern Indiana: A Study of the Upper Mississippi Cultures of the Kankakee Valley.* Indianapolis: Indiana Historical Society, Prehistory Research Series, Vol. 5, No. 1, July 1972.

Ford, Richard I. "The Processes of Plant Food Production in Prehistoric North America." In *Prehistoric Food Production in North America*, ed. Richard I. Ford. Ann Arbor: Museum of Anthropology, University of Michigan, Anthropology Paper No. 75, 1985.

Foster, Emily, ed. *American Grit: A Woman's Letters from the Ohio Frontier.* Lexington: University Press of Kentucky, 2002.

Freyfogel, Eric T. *Bounded People, Boundless Lands: Envisioning a New Land Ethic.* Washington, DC: Island Press, 1998.

Fuller, Margaret. *Summer on the Lakes, in 1843.* Boston: Charles C. Little and James Brown, 1844.

Furlong, Patrick J. "Plowmakers for the World." *Traces of Indiana and Midwestern History* 10, no. 2 (Spring 1998): 36–39.

Galinat, Walton C. "Domestication and Diffusion of Maize." In *Prehistoric Food Production in North America*, ed. Richard I. Ford. Ann Arbor: Museum of Anthropology, University of Michigan, Anthropology Paper No. 75, 1985.

———. "The Evolution of Corn and Culture in North America." In *Prehistoric Agriculture*, ed. Stuart Struever. Gargen City, NY: Natural History Press, 1971.

Gates, Paul W. *Agriculture and the Civil War.* New York: Knopf, 1965.

———. *The Farmer's Age: Agriculture, 1815–1860.* New York: Holt, Rinehart and Winston, 1960.

———. "Hoosier Cattle Kings." *Indiana Magazine of History* 44 (1948): 1–24.

———. *Landlords and Tenants on the Prairie Frontier: Studies in American Land Policy.* Ithaca: Cornell University Press, 1973.

Gipson, Lawrence Henry (ed.). T*he Moravian Indian Mission on White River: Diaries and Letters, May 5, 1799 to November 12, 1806.* Translated from German by Harry E. Stocker, et al. Indianapolis: Indiana Historical Bureau, 1938.

Gitlin, Jay. "On the Boundaries of Empire: Connecting the West to the Imperial Past." In *Under an Open Sky: Rethinking America's Western Past*, ed. William Cronon, et al. New York: Norton, 1992.

Glenn, Elizabeth, and Stewart Rafert. "Native Americans." In *Peopling Indiana: The*

Ethnic Experience, ed. Robert M. Taylor, Jr. and Connie A. McBirney. Indianapolis: Indiana Historical Society, 1996.

Goodman, Paul. "The Emergence of Homestead Exemption in the United States: Accomodation and Resistance to the Market Revolution." *Journal of American History* 80 (1993): 470–498.

Gregson, Mary Eschelbach. "Rural Response to Increased Demand: Crop Choice in the Midwest, 1860–1880." *Journal of Economic History* 53 (1993): 332–345.

Handbook of North American Indians. Vol. 15: *Northeast*, ed. Bruce G. Trigger. Washington: Smithsonian Institution, 1978.

Harper, Douglas. *Changing Works: Visions of a Lost Agriculture*. Chicago: University of Chicago Press, 2001.

Harper, R. Eugene. *The Transformation of Western Pennsylvania, 1770–1800*. Pittsburgh: University of Pittsburgh Press, 1991.

Hastorf, Christine A., and Sissel Johannessen. "Becoming Corn-Eaters in Prehistoric America." In *Corn and Culture in the Prehistoric New World*, ed. Sissel Johannessen and Christine A. Hastorf. Boulder, CO: Westview Press, 1994.

Headlee, Sue. *The Political Economy of the Family Farm: The Agrarian Roots of American Capitalism*. New York: Praeger, 1991.

Henlein, Paul C. *Cattle Kingdom in the Ohio Valley*. Lexington: University of Kentucky Press, 1959.

Higgs, Robert. *The Transformation of the American Economy, 1865–1914: An Essay in Interpretation*. New York: John Wiley & Sons, 1971.

Hill, Leonard U. *John Johnston and the Indians in the Land of the Three Miamis*. Piqua, OH: n.p., 1957.

Hinderaker, Eric. *Elusive Empires: Constructing Colonialism in the Ohio Valley, 1673–1800*. New York: Cambridge University Press, 1997.

History of Switzerland County, Indiana. Chicago: Weakley, Harraman, & Co., Publishers, 1885.

Hodge, Frederick Webb (ed.). *Handbook of American Indians North of Mexico*. 2 parts. Washington: Government Printing Office, 1905.

Hofstra, Warren. "'The Extension of His Majesties Dominions': The Virginia Backcountry and the Reconfiguration of Imperial Frontiers." *Journal of American History* 84 (1998): 1281–1312.

———. "Land Policy and Settlement in the Northern Shenandoah Valley." In *Appalachian Frontiers*, ed. Robert D. Mitchell. Lexington: University Press of Kentucky, 1991.

———. *The Planting of New Virginia: Settlement and Landscapes in the Shenandoah Valley*. Baltimore: Johns Hopkins University Press, 2004.

"Hoosier History: Down on the Farm." 1996. Floyd County Museum, project funded by an Indiana Heritage Research Grant. Accessed April 8, 1998 at <http://www.inc4u.org/farm.htm>.

Houk, Howard Jacob. "A Century of Indiana Farm Prices, 1841 to 1941." Lafayette,

IN: Purdue University, Agricultural Experiment Station, Bulletin No. 476, January 1943.

Howe, Daniel Wait. *A Descriptive Catalogue of the Official Publications of the Territory and State of Indiana from 1800 to 1890*. In *Indiana Historical Society Publications*, Vol. 2, No. 5. Indianapolis: Bowen-Merrill Company, 1890.

Hudson, John C. *Making the Corn Belt: A Geographical History of Middle-Western Agriculture*. Bloomington: Indiana University Press, 1994.

Hughes, Jonathan. *American Economic History*, expanded edition. Glenview, IL: Scott, Foresman, 1987.

———. "The Great Land Ordinances: Colonial America's Thumbprint on History." In *Essays on the Economy of the Old Northwest*, ed. David C. Klingaman and Richard K. Vedder. Athens: Ohio University Press, 1987.

Hurt, R. Douglas. *American Farm Tools: From Hand-Power to Steam-Power*. Manhattan, KS: Sunflower University Press, 1982.

———. *Indian Agriculture in America: Prehistory to the Present*. Lawrence: University Press of Kansas, 1987.

———. "Midwestern Distinctiveness." In *The American Midwest: Essays on Regional History*, ed. Andrew R. L. Cayton and Susan E. Gray. Bloomington: Indiana University Press, 2001.

Illustrated Historical Atlas of the State of Indiana. Chicago, 1870; partially reprinted Indianapolis: Indiana Historical Society, 1968.

Jager, Ronald. *The Fate of Family Farming: Variations on an American Idea*. Hanover, NH: University Press of New England, 2004.

———. *Eighty Acres: Elegy for a Family Farm*. Boston: Beacon Press, 1990.

Jakle, John A. *Images of the Ohio Valley: A Historical Geography of Travel, 1740-1860*. New York: Oxford University Press, 1977.

James, John A. *Money and Capital Markets in Postbellum America*. Princeton: Princeton University Press, 1978.

Jensen, Joan M. "Native American Women and Agriculture: A Seneca Case Study." In *Unequal Sisters: A Multicultural Reader in U.S. Women's History*, ed. Ellen Carol DuBois and Vicki L. Ruiz. New York: Routledge, 1990.

Johnson, Allen. "In Search of the Affluent Society." *Human Nature* (Sept. 1978), pp. 50–59.

Johnson, Oliver. *A Home in the Woods: Pioneer Life in Indiana: Oliver Johnson's Reminiscences of Early Marion County*, second edition. Bloomington: Indiana University Press, 1978.

Jones, Robert Leslie. *History of Agriculture in Ohio to 1880*. Kent, OH: Kent State University Press, 1983.

Kaplan, Lawrence. "Archeology and Domestication in American Phaseolus (Beans)." In *Prehistoric Agriculture*, ed. Stuart Struever. Garden City, NY: Natural History Press, 1971.

Kellar, Herbert Anthony (ed.). *Solon Robinson: Pioneer and Agriculturalist.* 2 vols. Indianapolis: Indiana Historical Bureau, 1936.

Kellar, James H. *An Introduction to the Prehistory of Indiana.* Indianapolis: Indiana Historical Society, 1973.

Killian, Larita J. "Mint Farming in Lakeville." *Traces of Indiana and Midwestern History* 16, no. 1 (Winter 2004): 44–47.

King, Francis B. "Early Cultivated Cucurbits in Eastern North America." In *Prehistoric Food Production in North America*, ed. Richard I. Ford. Ann Arbor: Museum of Anthropology, University of Michigan, Anthropology Paper No. 75, 1985.

Kirby, Jack Temple. "Rural Culture in the American Middle West: Jefferson to Jane Smiley." *Agricultural History* 70 (1996): 581–597.

Knodell, Jane. "The Demise of Central Banking and the Domestic Exchanges: Evidence from Antebellum Ohio." *Journal of Economic History* 58 (1998): 714–730.

Lambert, Joseph B. *Traces of the Past: Unraveling the Secrets of Archaeology through Chemistry.* Reading, MA: Addison-Wesley, 1997.

Lanier, James Franklin Doughty. *Sketch of the Life of J.F.D. Lanier.* New York: Hosford and Sons, Printers, 1877.

Larson, John Lauritz, and David G. Vandersteel. "Agent of Empire: William Conner on the Indiana Frontier, 1800–1855." *Indiana Magazine of History* 80 (1984): 301–328.

Latta, William C. *Outline History of Indiana Agriculture.* Lafayette, IN: Purdue University, Agricultural Experiment Station/Department of Agricultural Extension, 1938.

Lebergott, Stanley. "'O Pioneers': Land Speculation and the Growth of the Midwest." In *Essays on the Economy of the Old Northwest*, ed. David C. Klingaman and Richard K. Vedder. Athens: Ohio University Press, 1987.

Lincoln, Abraham. *Complete Works of Abraham Lincoln*, new and enlarged edition. Ed. John G. Nicolay and John Hay. Vol. 7. Harrowgate, TN: Lincoln Memorial University, 1894.

Lindley, Harlow (ed.). *Indiana as Seen by Early Travelers.* Indianapolis: Indiana Historical Commission, 1916.

Lyon, Bryce. *The Origins of the Middle Ages: Pirenne's Challenge to Gibbon.* New York: W.W. Norton, 1970.

MacMaster, Richard K. "The Cattle Trade in Western Virginia, 1760–1830," in *Appalachian Frontiers*, ed. Robert D. Mitchell. Lexington: University Press of Kentucky, 1991.

Madison, James D. *The Indiana Way: A State History.* Bloomington: Indiana University Press, 1986.

Marks, Robert B. *The Origins of the Modern World: A Global and Ecological Narrative.* Lanham, MD: Rowman & Littlefield, 2002.

Martin, Calvin. *Keepers of the Game: Indian-Animal Relationships and the Fur Trade*. Berkeley: University of California Press, 1978.

———. "The Metaphysics of Writing Indian-White History." In *The American Indian and the Problem of History*, ed. Calvin Martin. New York: Oxford University Press, 1987.

Martin, Marshall A. "The Economic Transformation of Indiana Agriculture: An Historic Perspective." *Proceedings of the Indiana Academy of the Social Sciences* 22 (1987): 32–39.

Mathews, Lois Kimball. *The Expansion of New England*. New York: Russell & Russell, 1909.

McClellend, Peter D. *Sowing Modernity: America's First Agricultural Revolution*. Ithaca: Cornell University Press, 1997.

McCord, Shirley S. (compiler). *Travel Accounts of Indiana, 1679–1961*. Indianapolis: Indiana Historical Bureau, 1970.

McCutcheon, Marc. *Everyday Life in the 1800s: A Guide for Writers, Students and Historians*. Cincinnati: Writer's Digest Books, 1993.

McIsaac, Gregory. "Sustainable Agriculture: Weaving the Pieces Together into a Coherent System." In *Sustainable Agriculture in the American Midwest: Lessons from the Past, Prospects for the Future*, ed. Gregory McIsaac and William R. Edwards. Urbana: University of Illinois Press, 1994.

McMurry, Sally. "Progressive Farm Families and Their Houses, 1830–1855: A Study in Independent Design." *Agricultural History* 58 (1984): 330–346.

Meyer, David R. "Emergence of the American Manufacturing Belt: An Interpretation." *Journal of Historical Geography* 9 (1983): 145–174.

Michaux, Phyllis (ed.). "Instructions from the Ohio Valley to French Emigrants." *Indiana Magazine of History* 84 (1988): 161–175.

Miller, Leonora Paxton. "Doctors, Drugs, and Disease in Pioneer Princeton." *Indiana Magazine of History* 52 (1956), pp. 141–156.

Mills, Stephanie. *Robert Swann*. Grabiola Island, B.C.: New Society Publishers, forthcoming..

Modell, John. "Family and Fertility on the Indiana Frontier, 1820." *American Quarterly* 23, no. 5 (Dec. 1971): 615–634.

Morrison, Olin Dee. *Indiana: "Hoosier State,"* 3 vols. Mimeographed. Athens, OH: E. M. Morrison, 1958.

Nation, Richard F. *At Home in the Hoosier Hills: Agriculture, Politics, and Religion in Southern Indiana, 1810–1870*. Bloomington: Indiana University Press, 2005.

Neth, Mary. *Preserving the Family Farm: Women, Community, and the Foundations of Agribusiness in the Midwest, 1900–1940*. Baltimore: Johns Hopkins University Press, 1995.

Nicholson, Howard Lee. "Swine, Timber, and Tourism: The Evolution of an Appalachian Community in the Middle West, 1830–1930." Ph.D. diss., Miami University, 1992.

North, Douglass C. *Growth and Welfare in the American Past*. Englewood Cliffs, NJ: Prentice-Hall, 1966.

Oberly, James W. *Sixty Million Acres: American Veterans and the Public Lands before the Civil War*. Kent, OH: Kent State University Press, 1990.

Paré, George. "The St. Joseph Mission." *Mississippi Valley Historical Review* 17, no. 1 (June 1930): 24–54.

Parker, William N. *Europe, America, and the Wider World: Essays on the Economic History of Western Capitalism*. 2 vols. Cambridge, Eng.: Cambridge University Press, 1991.

———. "From Northwest to Midwest: Social Bases of a Regional History." In *Essays in Nineteenth Century EconomicHistory: The Old Northwest*, ed. David C. Klingaman and Richard K. Vedder. Athens: Ohio University Press, 1975.

———. "Native Origins of Modern Industry." In *Essays on the Economy of the Old Northwest*, ed. David C. Klingaman and Richard K. Vedder. Athens: Ohio University Press, 1987.

Perkins, Elizabeth A. *Border Life: Experience and Memory in the Revolutionary Ohio Valley*. Chapel Hill: University of North Carolina Press, 1998.

Phillips, Clifton J. *Indiana in Transition: The Emergence of an Industrial Commonwealth, 1880–1920*. Indianapolis: Indiana Historical Bureau and Indiana Historical Society, 1968.

Pitzer, Donald E., and Josephine M. Elliott. "New Harmony's First Utopians, 1814-1824." *Indiana Magazine of History* 75 (1979): 225–300.

Post, Charles. "The 'Agricultural Revolution' in the United States: The Development of Capitalism and the Adoption of the Reaper in the Antebellum U.S. North." *Science and Society* 61 (1997): 216–228.

Power, Richard Lyle. *Planting Corn Belt Culture: The Impress of the Upland Southerner and Yankee in the Old Northwest*. Indianapolis: Indiana Historical Society, 1953.

Price, Robert. *Johnny Appleseed: Man and Myth*. 1954; reprinted Glochester, MA: Peter Smith, 1967.

Prices of Indiana Farm Products, 1841–1955. Station Bulletin 644 (1957). Purdue University, Agricultural Experiment Station, Lafayette, IN. [This work extends to 1955 the price data collected by Howard Jacob Houk through the year 1941.]

Prince, Hugh. *Wetlands of the American Midwest: A Historical Geography of Changing Attitudes*. Chicago: University of Chicago Press, 1997.

Pringle, Heather. "The Slow Birth of Agriculture." *Science* 282 (20 November 1999): 1446–1450.

Pudup, Mary Beth. "From Farm to Factory: Structuring and Location of the U.S. Farm Machinery Industry." *Economic Geography* 63 (1987): 203–222.

Randall, Willard Stern, and Nancy Nahra. *American Lives*, 2 vols. Vol. 1. New York: Longman, 1996.

Rankin, John. *Abolitionist: The Life of John Rankin, Written by Himself in His 80th Year*. Huntington, WV: Appalachian Movement Press, 1978.

Rasmussen, Wayne D., ed. *Readings in the History of American Agriculture*. Urbana: University of Illinois Press, 1960.

Rastatter, Edmund H. "Nineteenth Century Public Land Policy: The Case for the Speculator." In *Essays in Nineteenth Century Economic History: The Old Northwest*, ed. David C. Klingaman and Richard K. Vedder. Athens: Ohio University Press, 1975.

Redmond, Brian G. (ed.). *Current Research in Indiana Archaeology and Prehistory: 1991 and 1992*. Bloomington: Indiana University, Glenn A. Black Laboratory of Archaeology, Research Report No. 14, 1993.

Reidhead, Van A. *A Linear Programming Model of Prehistoric Subsistence Optimization: A Southeastern Indiana Example*. Indianapolis: Indiana Historical Society, Prehistory Research Series, Vol. 6, No. 1, 1981.

Rikoon, J. Stanford. *Threshing in the Midwest, 1820–1940: A Study of Traditional Culture and Technological Change*. Bloomington: Indiana University Press, 1988.

Roberts, Isaac Phillips. *Autobiography of a Farm Boy*. 1916; Ithaca: Cornell University Press, 1946.

Robertson, Robert S. *Valley of the Upper Maumee River*. Madison, WI: Brant & Fuller, 1889.

Robinson, Solon. See Kellar, Herbert Anthony (ed.).

Rockoff, Hugh. *The Free Banking Era: A Re-examination*. New York: Arno Press, 1975.

Rodgers, Thomas E. "Hoosier Women and the Civil War Home Front." *Indiana Magazine of History* 97 (2001): 105–128.

Rohrbough, Malcomb J. "Diversity and Unity in the Old Northwest, 1790–1840: Several Peoples Fashion a Single Region." In *Pathways to the Old Northwest*. Indianapolis: Indiana Historical Society, 1988.

———. *The Land Office Business: The Settlement and Administration of American Public Lands*. New York: Oxford University Press, 1968.

———. *The Trans-Appalachian Frontier: People, Societies, and Institutions, 1775–1850*. New York: Oxford University Press, 1978.

Roosevelt, Theodore. *The Winning of the West*. 2 vols. New York: Knickerbocker Press, 1889; reprinted New York: G.P. Putnam's Sons, 1927.

Rose, Gregory S. "Hoosier Origins: The Nativity of Indiana's United States-born Population in 1850." *Indiana Magazine of History* 81 (1985): 201–232.

———. "Upland Southerners: The County Origins of Southern Migrants to Indiana by 1850." *Indiana Magazine of History* 82 (1986): 242–263.

Rothstein, Morton. *Writing American Agricultural History*. Marshall, MN: Dept. of History, Southwest State University, 1996.

Rugh, Susan Sessions. *Our Common Country: Family Farming, Culture, and Community in the Nineteenth-Century Midwest*. Bloomington: Indiana University Press, 2001.

Rutherford, Justine Felix. *Wild Mustard: Flavorful Characters, Opinions, and Recipes from Spurlock Creek*. Huntington, WV: Mid-Atlantic Highlands Publishing, 2005.

Sahlins, Marshall. "The Original Affluent Society." In Marshall Sahlins, *Stone Age Economics*. Chicago: Aldine-Atherton, 1972.

Salamon, Sonya. *Prairie Patrimony: Family, Farming, and Community in the Midwest*. Chapel Hill: University of North Carolina Press, 1992.

Salisbury, Neal. "American Indians and American History." In *The American Indian and the Problem of History*, ed. Calvin Martin. New York: Oxford University Press, 1987.

Sauer, Carl. "Homestead and Community on the Middle Border." In Carl Sauer, *Land and Life*. Berkeley: University of California Press, 1969.

Schapsmeier, Edward L., and Frederick H. Schapsmeier. *Encyclopedia of American Agricultural History*. Westport, CT: Greenwood Press, 1975.

Schery, Robert W. *Plants for Man*. New York: Prentice-Hall, 1952.

Schob, David E. *Hired Hands and Plowboys: Farm Labor in the Midwest, 1815–60*. Urbana: University of Illinois Press, 1975.

Schramm, Jacob. *The Schramm Letters: Written by Jacob Schramm and Members of His Family from Indiana to Germany in the Year 1836*, trans. and ed. Emma S. Vonnegut. Indianapolis: Indiana Historical Society, 1936.

Schumacher, E. F. *Small Is Beautiful: Economics As If People Mattered*. New York: Harper & Row, 1973.

Schwartzweller, Harry K., James S. Brown, and J. J. Mangalam. *Mountain Families in Transition: A Case Study of Appalachian Migration*. University Park: Pennsylvania State University Press, 1971.

Scott, Donald H. *Barns of Indiana*. Virginia Beach, VA: The Donning Company, Publishers, 1997.

Sennett, Richard. *Families Against the City: Middle Class Homes of Industrial Chicago, 1872–1890*. Cambridge: Harvard University Press, 1970.

Shaffer, Lynda Norene. *Native Americans before 1492: The Moundbuilding Centers of the Eastern Woodlands*. Armonk, NY: M.E. Sharpe, 1992.

Shankman, Andrew. *Crucible of American Democracy: The Struggle to Fuse Egalitarianism and Capitalism in Jeffersonian Pennsylvania*. Lawrence: University Press of Kansas, 2004.

Shannon, Fred A. *The Farmer's Last Frontier: Agriculture, 1860–1897*. New York: Holt, Rinehart and Winston, 1966.

Shaw, Albert. "The Progress of the World" [column]. *Review of Reviews* [New York] 92, no. 4 (October 1935): 11–18.

Shepard, Paul. Interview. In *Listening to the Land*, ed. Derrick Jensen. San Francisco: Sierra Club Books, 1996.

Sieber, Ellen, and Cheryl Ann Munson. *Looking at History: Indiana's Hoosier National Forest Region, 1600 to 1950*. 1992; Bloomington: Indiana University Press, 1994.

Smith, Bruce D. *The Emergence of Agriculture*, paperback edition. New York: Scientific American Library, 1998.

Smith, William Henry. *The History of the State of Indiana from the Earliest Explorations by the French to the Present Time*, 2 vols. Indianapolis: Western Publishing Company, 1903.

Somé, Malidoma Patrice. *Of Water and the Spirit: Ritual, Magic, and Initiation in the Life of an African Shaman*. New York: Putnam, 1994.

Stealey, John Edmund III. "Notes on the Antebellum Cattle Industry from the McNeill Family Papers." *Ohio History* 75 (Winter 1966): 38–47, 70–72.

Stoll, Steven. *Larding the Lean Earth: Soil and Society in Nineteenth-Century America*. New York: Hill and Wang, 2002.

Stout, David B. "Report on the Kickapoo, Illinois, and Potawatomi Indians." In *Indians of Illinois and Northwestern Indiana*, ed. David Agee Horr. New York: Garland, 1974.

Sturm, Philip W. "The Frontier." In *The West Virginia Encyclopedia*, ed. Ken Sullivan. Charleston, WV: West Virginia Humanities Council, 2006.

Sugden, John. *Tecumseh: A Life*. New York: Henry Holt, 1997.

Swierenga, Robert P. "The New Rural History: Defining the Parameters." *Great Plains Quarterly* 1 (1981): 211–223.

———. "Theoretical Perspectives on the New Rural History: From Environmentalism to Modernization." *Agricultural History* 56 (1982): 495–503.

Sydnor, Charles S. *The Development of Southern Sectionalism, 1819–1848*. Baton Rouge: Louisiana State University Press, 1948.

"Taking Action on Hog Farms." *Yes!* Magazine, issue 30 (Summer 2004): 5.

Taylor, George Rogers. *The Transportation Revolution, 1815–1860*. [Vol. 4 of Taylor's *Economic History of the United States*] New York: Holt, Rinehart, and Winston, 1951.

Taylor, Robert M., Jr. *The Northwest Ordinance, 1787: A Bicentennial Handbook*. Indianapolis: Indiana Historical Society, 1987.

Taylor, Robert M., Jr., and Connie A. McBirney, eds. *Peopling Indiana: The Ethnic Experience*. Indianapolis: Indiana Historical Society, 1996.

Thompson, Charles N. *Sons of the Wilderness: John and William Connor*. Indianapolis: Indiana Historical Society, 1937.

Thompson, Dave O., and William L. Madigan. *One Hundred and Fifty Years of Indiana Agriculture*. Indianapolis: Indiana Sesquicentennial Commission, 1966.

Thornbrough. Emma Lou. *Indiana in the Civil War Era, 1850–1880*. Indianapolis: Indiana Historical Bureau and Indiana Historical Society, 1965.

———. *The Negro in Indiana before 1900: A Study of a Minority*. Indianapolis: Indiana Historical Bureau, 1957; reprinted Bloomington: Indiana University Press, 1993.

Thorndale, William, and William Dollarhide. *Map Guide to the U.S. Federal Censuses, 1790–1920*. Baltimore: Genealogical Publishing, 1987.

Turner, Frederick Jackson. *The Frontier in American History*. New York: Henry Holt, 1921.

United States Census Bureau. Census of the United States, 1840–1870.

———. *Historical Statistics of the United States, Colonial Times to 1970*, 2 parts. Bicentennial edition. Washington, DC: Government Printing Office, 1975.

———. *Statistical Abstract of the United States, 2004–2005*. Washington, DC, 2004.

United States Dept. of Agriculture. *Soil Survey: Madison County, Indiana*. Washington, DC: Government Printing Office, March 1967.

Vance, Rupert. *Human Geography of the South: A Study in Regional Resources and Human Adequacy*. Chapel Hill, University of North Carolina Press, 1932.

Vedder, Richard K., and Lowell E. Gallaway. "Migration and the Old Northwest." In *Essays in Nineteenth Century Economic History: The Old Northwest,* ed. David C. Klingaman and Richard K. Vedder. Athens: Ohio University Press, 1975.

Vincent, Stephen A. *Southern Seed, Northern Soil: African-American Farm Communities in the Midwest, 1765–1900*. Bloomington: Indiana University Press, 1999.

Wagner, Gail E. "Corn in Eastern Woodlands Late Prehistory." In *Corn and Culture in the Prehistoric New World*, ed. Sissel Johannessen and Christine A. Hastorf. Boulder, CO: Westview Press, 1994.

Walsh, Margaret. *The Rise of the Midwestern Meat Packing Industry*. Lexington: University Press of Kentucky, 1982.

Walton, Gary M. "River Transportation and the Old Northwest Territory." In *Essays on the Economy of the Old Northwest*, ed. David C. Klingaman and Richard K. Vedder. Athens: Ohio University Press, 1987.

Warren, Dennis Michael. "Indigenous Agricultural Knowledge, Technology, and Social Change." In *Sustainable Agriculture in the American Midwest: Lessons from the Past, Prospects for the Future*, ed. Gregory McIsaac and William R. Edwards. Urbana: University of Illinois Press, 1994.

Warren, Louis A. *Lincoln's Youth: Indiana Years, Seven to Twenty-one, 1816–1830*. Indianapolis: Indiana Historical Society, 1959.

Weber, Max. *The Agrarian Sociology of Ancient Civilizations*, trans. by R. I. Frank. London: NLB, 1976.

Wessel, Thomas R. "Agriculture, Indians, and American History." *Agricultural History* 50 (1976): 9–20.

Whitaker, James W. *Feedlot Empire: Beef Cattle Feeding in Illinois and Iowa, 1840–1900*. Ames: Iowa State University Press, 1975.

Wilkinson, William C., ed. "'To do for my self': Footloose on the Old Northwest Frontier." *Indiana Magazine of History* 86 (1990): 399–420.

Williams, William Appleman. *The Roots of the Modern American Empire*. New York: Random House, 1969.

Wilson, Denise Marie. "Vincennes: From French Colonial Village to American Frontier Town, 1730–1820." Ph.D. diss., West Virginia University, 1997.

Winquist, Alan H. "Scandinavians." In *Peopling Indiana: The Ethnic Experience*, ed. Robert M. Taylor, Jr. and Connie M. McBirney. Indianapolis: Indiana Historical Society, 1996.

Winter, George. *Indians and a Changing Frontier: The Art of George Winter*. Indianapolis: Indiana Historical Society, 1993.

Witthoft, John. *The American Indian as Hunter*. Harrisburg: Pennsylvania Historical and Museum Commission, 1990.

Woehrmann, Paul. *At the Headwaters of the Maumee: A History of the Forts of Fort Wayne*. Indianapolis: Indiana Historical Society, 1971.

Woods, Sam B. *The First Hundred Years of Lake County, Indiana*. [Crown Point, IN?], 1938.

Yale Bulletin Calendar, 23 June 2000. Accessed 25 July 2005 at www.econ.yale.edu/news/parker/parker.htm

Yarnell, Richard A. "Early Plant Husbandry in Eastern North America." In *Cultural Change and Continuity: Essays in Honor of James Bennett Griffin*, ed. Charles E. Cleland. New York: Academic Press, 1976.

———. *Aboriginal Relationships between Culture and Plant Life in the Upper Great Lakes Region*. Ann Arbor: Museum of Anthropology, University of Michigan, Anthropology Paper No. 23, 1964.

Index